If we forecast a weather event, what effect will this have on people's behavior, and how will their change in behavior influence the economy? We know that weather and climate variations can have a significant impact on the economics of an area, and just how weather and climate forecasts can be used to mitigate this impact is the focus of this book.

Adopting the viewpoint that information about the weather has value only insofar as it affects human behavior, contributions from economists, psychologists, and statisticians, as well as meteoreologists, provide a comprehensive view of this timely topic. These contributions encompass forecasts over a wide range of temporal scales, from the weather over the next few hours to the climate months or seasons ahead.

Economic Value of Weather and Climate Forecasts seeks to determine the economic benefits of existing weather forecasting systems and the incremental benefits of improving these systems, and will be an interesting and essential text for economists, statisticians, and meteorologists.

ECONOMIC VALUE OF
WEATHER AND CLIMATE FORECASTS

ECONOMIC VALUE OF
WEATHER AND CLIMATE FORECASTS

Edited by

RICHARD W. KATZ
National Center for Atmospheric Research, USA

ALLAN H. MURPHY
Prediction and Evaluation Systems, USA

CAMBRIDGE
UNIVERSITY PRESS

PUBLISHED BY THE PRESS SYNDICATE OF THE UNIVERSITY OF CAMBRIDGE
The Pitt Building, Trumpington Street, Cambridge CB2 1RP, United Kingdom

CAMBRIDGE UNIVERSITY PRESS
The Edinburgh Building, Cambridge CB2 2RU, United Kingdom
40 West 20th Street, New York, NY 10011-4211, USA
10 Stamford Road, Oakleigh, Melbourne 3166, Australia

First published 1997

Typeset in Times Roman

Library of Congress Cataloging-in-Publication Data

Economic value of weather and climate forecasts / edited by Richard W.
Katz, Allan Murphy.
p.   cm.
Includes index.
ISBN 0-521-43420-3
1. Weather forecasting – Economic aspects. 2. Long-range weather
forecasts – Economic aspects.   I. Katz, Richard W.   II. Murphy,
Allan H.
QC995.E29   1997
338.4'755163 – dc21   96-50389

A catalog record for this book is available from
the British Library

ISBN 0-521-43420-3 hardback

Transferred to digital printing 2002

Contents

Preface

The topic of this book brings to mind an oft-quoted adage: everyone talks about the weather, but no one does anything about it. Despite this adage, the focus of this volume is not the weather itself, or weather forecasting per se, or even the various economic impacts of weather, but rather the way in which weather forecasts can be utilized to mitigate these impacts. The viewpoint adopted here is that information about the weather has value only insofar as it affects human behavior. Despite their inherent imperfections, weather forecasts have the potential to influence behavior. To draw an analogy, even quite small but real shifts in the odds can produce attractive returns when playing games of chance.

It is indeed true that "talk" about the weather abounds. Relatively large expenditures are devoted to both observational systems and research programs intended to enhance weather forecasting capability, as well as to operational activities related to the production and distribution of forecasts to a variety of users. Moreover, many of the substantial economic impacts of various weather events are well documented. Somewhat surprisingly, however, relatively little attention has been devoted to determining the economic benefits of existing weather forecasting systems or the incremental benefits of improvements in such systems.

This lack of attention may partly reflect the fact that assessing the economic value of weather forecasts is a challenging problem; among other things, it is an inherently multidisciplinary endeavor. Besides the field of meteorology, the disciplines include economics (a monetary value is attached to a publicly available good), psychology (human behavior under uncertainty influences forecast use and value), and statistics as well as closely related fields of management science and operations research (the formal assessment process utilizes the principles of decision theory). All these disciplines are represented in the backgrounds of the contributors to the present volume.

The scope of the book encompasses forecasts over a wide range of temporal scales. Included are relatively long-range (e.g., monthly or seasonal) predictions, sometimes referred to as "climate forecasts." This term should not be confused with "climate change," a topic that is not covered here, in part because operational predictions of climate change are not yet produced on a regular basis. In view of the new long-lead climate outlooks produced by the U.S. National Weather Service, as well as the recently reinvigorated U.S. Weather Research Program, whose ultimate goal is to improve short-range weather forecasts, a book on the economic value of forecasts appears especially timely. It could even be argued that weather forecasts themselves constitute a resource of general interest to researchers concerned with decision making under uncertainty. After all, few other forecasting systems come to mind in which the predictions are routinely made available to a broad spectrum of potential users and in which it is possible to evaluate forecasting performance in a relatively straightforward and timely manner.

Chapter 1, "Weather Prediction," by Joseph J. Tribbia, describes the scientific basis of modern weather forecasting, with emphasis on the so-called numerical (i.e., physical–dynamical) component of the forecasting process. The highly nonlinear mathematical equations governing the time evolution of the state of atmosphere are presented. Moreover, the worldwide network of meteorological observations is such that this state is only incompletely observed at any given time. These two factors combine to produce the phenomenon of chaos, thereby limiting the predictability of day-to-day weather conditions. Brief reference is also made to the numerical-statistical procedures currently used to produce routine forecasts of surface weather conditions.

Chapter 2, "Forecast Verification," by Allan H. Murphy, describes an approach to forecast evaluation that recognizes the fundamental role of the joint distribution of forecasts and observations in the verification process, and focuses on a suite of methods designed to measure the various attributes of forecast quality. In addition, the concept of "sufficiency" is introduced as a means of screening two or more competing weather forecasting systems. Only when the sufficiency relation can be shown to hold between pairs of systems can it be unambiguously stated that one system dominates the other, in terms of being of at least as much value to all users.

Chapter 3, "The Value of Weather Information," by Stanley R. Johnson and Matthew T. Holt, presents the fundamental tenets of Bayesian decision theory, in which the criterion of selecting the action that maximizes expected utility is adopted. Being normative in nature, this theory prescribes how individual decision makers ought to employ imperfect weather forecasts. Determining the economic value of an imperfect weather forecasting system entails a comparison of the expected utility with and without the system. In the absence of any forecasts, it is often reasonable to assume that the decision maker has access to historical probabilities of weather events, termed "climatological information." Also treated are other economic issues, including methods of determining the value of a forecasting system to society as a whole.

Chapter 4, "Forecast Value: Prescriptive Decision Studies," by Daniel S. Wilks, reviews case studies that have adopted the normative/prescriptive approach introduced in Chapter 3. The vast majority of such studies involve agriculture; other areas of application include forestry and transportation. Some of the more realistic case studies have required the modeling of sequential decision-making problems, that is, dynamic situations in which the action taken and the event that occurs at the present stage of the problem are related to actions and events at subsequent stages. At least in limited circumstances, these studies establish both that present forecasting systems can have substantial value and that nonnegligible incremental benefits could be realized with hypothetical improvements in such systems.

Chapter 5, "Forecast Value: Descriptive Decision Studies," by Thomas R. Stewart, reviews how individual users of weather forecasts actually behave in response to those forecasts. Research on judgment and decision making conducted by cognitive psychologists reveals that individuals do not necessarily behave in a manner consistent with the principle of maximizing expected utility, on which the prescriptive approach is predicated. Descriptive studies of the use of weather forecasts range from simple surveys to detailed monitoring of decision makers in action. Unfortunately, descriptive studies to date have lacked sufficient detail to produce actual estimates of the value of weather forecasts. Ultimately, the descriptive information provided serves to complement and inform prescriptive case studies such as those covered in Chapter 4.

Chapter 6, "Forecast Value: Prototype Decision-Making Models," by Richard W. Katz and Allan H. Murphy, utilizes the sufficiency concept introduced in Chapter 2 as well as the normative methodology described in Chapter 3. Prototype decision-making

models are treated, ranging from the simplest case of a static
decision-making problem — such as whether or not to carry an
umbrella in response to uncertainty about the occurrence of rain
— to more complex, dynamic problems that mimic some of the
essential features of real-world case studies reviewed in Chapter
4. For such prototype models, analytical results are derived concerning how economic value increases as a function of the quality
of the forecasting system. These results include the existence of a
threshold in forecast quality below which economic value is zero
and above which value increases as a convex function (i.e., its slope
is also an increasing function) of quality.

We thank Barbara Brown, Andrew Crook, William Easterling,
Roman Krzysztofowicz, Kathleen Miller, and especially Martin
Ehrendorfer for serving as reviewers of individual chapters. We
are also grateful for the assistance in manuscript preparation provided by Jan Hopper, Maria Krenz, Jan Stewart, and the late
Shirley Broach. The National Center for Atmospheric Research is
operated by the University Corporation for Atmospheric Research
under sponsorship of the National Science Foundation.

Richard W. Katz
Boulder, Colorado

Allan H. Murphy
Corvallis, Oregon

Contributors

MATTHEW T. HOLT, associate professor of agricultural and resource economics at North Carolina State University, received a Ph.D. in agricultural economics from the University of Missouri. Previously, Dr. Holt was employed by Iowa State University and the University of Wisconsin. His current research interests include the role of risk and information in agricultural production and resource allocation decisions, and the role of dynamics in managed populations and agricultural markets. His publications include work on rational expectations modeling, the role of risk and uncertainty in the presence of government price support and supply management interventions, and the potential for nonlinear dynamics in agricultural markets.

STANLEY R. JOHNSON is C.F. Curtiss Distinguished Professor in agriculture and director of the Center for Agricultural and Rural Development (CARD), department of economics, Iowa State University. Previously, Dr. Johnson was employed at the University of Missouri, University of California–Berkeley, Purdue University, University of California–Davis, University of Georgia, and University of Connecticut. His related interests are in agriculture sector and trade policy, food and nutrition policy, and natural resources and environmental policy. His work prior to and at CARD has emphasized analysis of policy processes and the use of analytical systems to evaluate policy options. He has authored the following books: *Advanced Econometric Methods, Demand Systems Estimation: Methods and Applications*, and *Agricultural Sector Models for the United States: Descriptions and Selected Policy Applications*. He has co-authored several books, of which *Conservation of Great Plains Ecosystems: Current Science, Future Options* is the most recent.

RICHARD W. KATZ, senior scientist and deputy head of the Environmental and Societal Impacts Group at the National Center for Atmospheric Research (NCAR), received a Ph.D. in statistics from Pennsylvania State University. Previously, Dr. Katz was employed by Oregon State University and the National Oceanic and Atmospheric Administration. His research interests focus on the application of probability and statistics to atmospheric and related sciences and to assessing the societal impact of weather and climate. He has been active in promoting multidisciplinary research, especially through collaboration between statistical and atmospheric scientists. His publications include the co-edited books *Teleconnections Linking Worldwide Climate Anomalies* and *Probability, Statistics, and Decision Making in the Atmospheric Sciences*. NCAR is sponsored by the National Science Foundation.

ALLAN H. MURPHY is a principal of Prediction and Evaluation Systems (Corvallis, Oregon) and professor emeritus at Oregon State University. He was awarded M.A. and Ph.D degrees in mathematical statistics and atmospheric sciences, respectively, by the University of Michigan. Previously, Dr. Murphy was employed at the National Center for Atmospheric Research, the University of Michigan, and Travelers Research Center (Hartford, Connecticut). He has held visiting appointments at various universities and research institutes in the United States, Europe, and Asia. His research interests focus on the application of probability and statistics to atmospheric sciences, with particular emphasis on probability forecasting, forecast verification, and the use and value of forecasts. Dr. Murphy's publications include approximately 150 papers in the refereed literature across several fields. His co-edited volumes include *Weather Forecasting and Weather Forecasts: Models, Systems, and Users* and *Probability, Statistics, and Decision Making in the Atmospheric Sciences*.

THOMAS R. STEWART, director for research, Center for Policy Research, University at Albany, State University of New York, received a Ph.D. in psychology from the University of Illinois. Previously, Dr. Stewart was employed at the Graduate School of Public Affairs and the Center for Research on Judgment and Policy at the University of Colorado and the Environmental and Societal Impacts Group at the National Center for Atmospheric Research. His research interests focus on the application of judgment and decision research to problems involving scientific and technical expertise and public policy, including studies of regional air quality policy, visual air quality judgments, use of weather forecasts in agriculture, risk analysis, scientists' judgments about global climate change, management of dynamic systems, and the judgments of expert weather forecasters.

JOSEPH J. TRIBBIA is senior scientist and head of the global dynamics section of the Climate and Global Dynamics Division at the National Center for Atmospheric Research (NCAR). He received a Ph.D. in atmospheric sciences from the University of Michigan. His work at NCAR has focused on the numerical simulation of the atmosphere and geophysically relevant flows. His research includes work on the application of dynamical systems theory in atmospheric dynamics, the problems of atmospheric data analysis and numerical weather prediction, and most recently the simulation and prediction of El Niño–Southern Oscillation. He serves as an editor of the *Journal of the Atmospheric Sciences*. NCAR is sponsored by the National Science Foundation.

DANIEL S. WILKS is associate professor in the department of soil, crop, and atmospheric sciences at Cornell University. He received a Ph.D. in atmospheric sciences from Oregon State University. His research involves primarily applications of probability and statistics to meteorological and climatological problems, and to weather- and climate-sensitive areas such as agriculture. He is author of the recently published textbook *Statistical Methods in the Atmospheric Sciences*.

vision of the National Center for Atmospheric Research (NCAR). He received his Ph.D. in atmospheric sciences from the University of Michigan. His work at NCAR has focused on the numerical simulation of the atmosphere and geophysical... flows. His research includes work on the application of dynamical data... the... in atmospheric dynamics, the problems of atmospheric data analysis and important weather prediction, and most recently the simulation and control of the spread of... air... pollution. He is an editor of the Journal of the Atmospheric Sciences... ... ... sponsored by the National Science Foundation.

DANIEL S. WILKS is associate professor in the department of soil, crop, and atmospheric sciences at Cornell University. He received his Ph.D. in atmospheric sciences from... State University. He teaches... in the structure... applications of statistics, and... statistics to meteorological and climatological variables and... the weather and climate... he teaches as well as... ... author of the recently published book Statistical Methods in the Atmospheric Sciences.

1

Weather prediction

JOSEPH J. TRIBBIA

1. History and introduction

The public tends to have certain misconceptions about the nature of research on weather prediction. It is not that the general populace is of the opinion that weather prediction is a field so developed and accurate that research is not necessary. Nor is it surprising to most that modern weather prediction makes heavy use of the most advanced computers available. Rather, it is the technique of prediction that is most unexpected to the nonspecialist. Most often it is supposed that a large, powerful computer is used to store an archive of past weather from which the most similar analog of the current weather is utilized to form a prediction of the future weather. While variants of such statistical techniques of prediction are still in use today for so-called extended-range predictions, the most accurate forecasts of short-range weather are based in large part on a deterministic application of the laws of physics. In this application, the computer is used to manipulate that vast amount of information needed to effect the solution of the equations corresponding to these physical laws with sufficient accuracy to be useful.

The recognition that the physical laws embodied in Newton's laws of motion and the first and second laws of thermodynamics could be applied to the problem of weather prediction has been attributed to Bjerknes (1904). It was only at the end of the nineteenth century that the application of Newton's laws to a compressible fluid with friction and the application of the empirical laws of the thermodynamics of an ideal gas could be joined to form a closed set of predictive equations with the same number of equations as unknowns. V. Bjerknes, a physicist interested in meteorology, recognized the relevance of these developments to the problem of weather forecasting. He laid out the prospect for weather prediction using the principal laws of science. Bjerknes and his collaborators embarked upon this path of prediction by using the following set of equations and variables:

(i) The Newtonian equations of motion relating the change of velocity of air parcels to the forces acting on such parcels; the gradient of pressure, gravity, frictional forces, and the fictitious Coriolis force, which is necessary when describing motion relative to a rotating reference frame such as the Earth:

$$\frac{d\mathbf{V}}{dt} + 2\mathbf{\Omega} \times \mathbf{V} = -\alpha\nabla p + \mathbf{g} + \mathbf{F}. \tag{1.1}$$

(ii) The equation of mass conservation or continuity equation, which reflects the impossibility of the spontaneous appearance of a vacuum in the atmosphere:

$$\frac{\partial \rho}{\partial t} + \nabla \cdot (\rho\mathbf{V}) = 0. \tag{1.2}$$

(iii) The equation of state for an ideal gas relating pressure, temperature, and density:

$$p\alpha = RT. \tag{1.3}$$

(iv) The first law of thermodynamics, reflecting the conservation of energy, including internal thermal energy of a gas:

$$C_v\frac{dT}{dt} + p\frac{d\alpha}{dt} = Q. \tag{1.4}$$

In equations (1.1) through (1.4), $\mathbf{V}$ is the three-dimensional velocity vector of the air; $\mathbf{\Omega}$ is the vector of the Earth's rotation pointing in the direction perpendicular to the surface at the North Pole; ∇ is the gradient operator; $\mathbf{g}$ is the vector acceleration due to gravity; $\mathbf{F}$ is the vector of frictional forces per unit mass; p is the atmospheric pressure; T is the atmospheric temperature; ρ is the atmospheric density; α is its inverse, specific volume; Q is the heating rate per unit mass of air; C_v is the specific heat at constant volume; R is the gas constant for dry air; and t is time. (For a fuller explanation and derivation of the above, see Petterssen, 1969.)

With the above predictive equations, Bjerknes stated that one could determine the future velocity, temperature, pressure, and density of the atmosphere from the present values of these quantities. That is to say, one could forecast the weather for indefinitely long periods of time. There was one difficulty, however, of which Bjerknes was aware, which is hidden in the forms of equations (1.1) through (1.4): When expanded for use in a geographically fixed reference frame (a so-called Eulerian perspective), these are non-linear equations, which precludes their solution in a closed form. Bjerknes intended to circumvent this difficulty by seeking graphical solutions.

The solution via graphical methods proved to be too cumbersome and inaccurate for Bjerknes to convince anyone of its utility. However, the theoretical concepts that Bjerknes brought to bear on the prediction problem did attract the interest of groups working on weather forecasting in Europe. In particular, the British took an interest in Bjerknes' scientific approach and sent a young weather observer, L. F. Richardson, to Bergen, Norway, to learn more about Bjerknes' ideas. Richardson was a former mathematics student of renown at Cambridge. He knew, through his own personal research, of a more accurate, more elegant, and simpler technique of solving systems of equations such as those above. This technique, which was ideally suited to the task of weather prediction, is called the finite difference method. Richardson began designing a grand test of the overall method of prediction and performing calculations in his spare time. World War I had begun and Richardson, a conscientious objector, found time to design his test during respites from his duties as an ambulance driver on the Western Front.

The story of Richardson's first grand test, which contains several twists of fate detailed in the account by Ashford (1985), would be impressive if the ensuing forecast had been successful. However, it was a complete disaster. Richardson predicted enormously large surface pressure changes in a six-hour forecast for one locale over Germany — the only location for which he computed a forecast. (A recent reconstruction of Richardson's calculation by Lynch [1994], using only slight modifications of Richardson's original method [which are obvious, given the perspective of modern-day knowledge], gave a very reasonable forecast.) Additionally, Richardson estimated that in order to produce computer-generated weather forecasts faster than the weather was changing, a group of 64,000 people working continuously with mechanical calculators and exchanging information was needed. Absorbed by his other interests, such as a mathematical theory on the development of international hostilities, Richardson never returned to the weather prediction problem. Despite the fact that his attempt was published in 1922 in a book entitled *Weather Prediction by Numerical Methods* (Richardson, 1922), the field lay dormant until the end of World War II.

Richardson's wildly inaccurate prediction highlighted the need for both scientific and technological advances to make weather prediction from scientific principles a useful endeavor. Both of these were forthcoming as indirect results of the war. The development of the electronic computer and the cooperative efforts in atmospheric observation, necessitated by military aviation, were precisely the advancements needed to bring to fruition scientifically based weather prediction using the governing physical laws. These advances were, of course, greatly buttressed by scientists' growing understanding of the atmosphere and its motions, knowledge gained over the twenty years during which research into physically based prediction was dormant. During that period, scientists including C. G. Rossby, J. G. Charney, E. T. Eady, R. C. Sutcliffe, J. Bjerknes, G. Holmboe, and A. Eliassen contributed their insights on how the atmosphere behaves, and they developed an understanding of its motions through the application of the physical laws enumerated above. Thus, when J. von Neumann envisioned scientific problems that the new electronic computer could address, weather prediction was a candidate for exploration.

In the mid 1940s, von Neumann met with the leading scientific lights of the time in meteorology to discuss the prospects for computer-produced weather prediction. The enthusiasm of Rossby, the most influential meteorologist in the world at that time, encouraged von Neumann to devote a portion of the scientific research at Princeton's Institute for Advanced Study to the numerical weather prediction project. A group of scientists — Charney and Eliassen, along with P. D. Thompson, R. Fjortoft, G. W. Platzman, N. A. Phillips, and J. Smagorinsky — set forth to produce a successful weather prediction from scientific laws and the Electronic Numerical Integrator and Computer (ENIAC).

In order to avoid the difficulty that thwarted Richardson's effort and to make the time necessary to produce a forecast as small as possible, the set of equations used to define the forecast was limited to a single equation that predicted the pressure at a single level approximately three miles up in the atmosphere. Even with this major simplification and limiting the domain of the forecast to the continental United States, the 24-hour forecast required 6 days of continuous computation to complete. This forecast was not only reasonable but also similar in accuracy to the subjective forecasts at the time. Thus the field now known as numerical weather prediction was reborn.

2. The modern era

From these beginnings, one might dare to predict a degree of success once the speed and power of computers caught up with the new understanding of atmospheric science. Indeed, an examination of the improvement of computer-produced weather forecasts since the mid 1950s, when such forecasts were operationally introduced, indicates that significant increases in skill occurred immediately following the availability of a new generation of computers and new numerical models.

Because of the rapid rise of computer power, the models used to predict the weather today are in many respects similar to the computational scheme developed by Richardson, with some elaborations. Richardson's forecast equations were in fact quite sophisticated, and many of the physical processes that Richardson included in his "model" of the atmosphere have only recently been incorporated into modern weather prediction models. For example, Richardson was the first to note the potential advantage of replacing the Newtonian relation between force and acceleration in the vertical direction with a diagnostic relationship between the force of gravity and the rate of change of pressure in the vertical. Thus Richardson replaced, and all modern (large-scale) weather prediction models replace, equation (1.1) above with:

$$\frac{d\mathbf{V}_h}{dt} + f\mathbf{k} \times \mathbf{V}_h = -\alpha\nabla_h p + \mathbf{F}_h, \qquad (1.1a)$$

$$\frac{\partial p}{\partial z} = -\rho\, g. \qquad (1.1b)$$

In equations (1.1a) and (1.1b), the subscript h denotes the horizontal component, f is the Coriolis parameter (i.e., the projection of the Earth's rotation in the direction perpendicular to the mean surface), $\mathbf{k}$ is a unit vector normal to the mean surface of the Earth, z is the coordinate in the vertical direction, and g is the scalar gravitational constant.

Richardson also included a prognostic equation for water vapor
in the atmosphere, which is a necessary component of present-day
forecast models:

$$\frac{\partial q}{\partial t} + \nabla \cdot (q\mathbf{V}) = S, \tag{1.5}$$

where q is the specific humidity (fractional mass of water vapor
in a unit mass of air) and S represents the sources and sinks of
water vapor such as evaporation and precipitation. The above set
of relationships (equations 1.1a, 1.1b, and 1.2–1.5) forms the basis
of most weather prediction models in existence today.

To delve further into the production of numerical weather predictions, it is necessary to explain in more detail the basic underlying
concepts in the transformation of the physical laws described above
into the arithmetic manipulations that Richardson envisioned of
64,000 employees and now performed at high speed by computer.
The equations above are formulated with regard to a conceptual
model of the atmosphere as a continuous, compressible fluid. For
the purpose of solving these equations in approximate form on
a computer, the continuous atmosphere must be subdivided into
manageable volumes that can be stored in a computer's memory.
Such a process is called "discretization"; one of the most common ways of discretizing the atmospheric equations is the finite
difference technique used by Richardson. In this discretization
technique, the atmosphere is subdivided into a three-dimensional
mesh of points. The averaged velocity, temperature, pressure, and
humidity for the volume of atmosphere surrounding each node on
this mesh are predicted using the physical equations (see Figure
1.1). Because the equations contain terms that require the derivative of the predicted quantities with respect to the spatial variables
(i.e., longitude, latitude, and height), these derivatives are approx-
imated by the difference of the quantity between neighboring grid
nodes divided by the distance between the nodes. Note that the
true derivative is simply the limit of this difference as the distance
between nodes approaches zero.

For current numerical weather prediction models, the distance
between grid nodes is between one and two degrees of longitude
or latitude (between 110 and 220 km at the equator) in the horizontal and between 500 m and 1 km in the vertical. The above
figures are for the models used for global operational prediction at

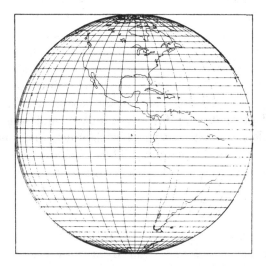

Figure 1.1. Example of grid lattice covering the earth. This grid consists of 40 nodes in the latitudinal direction and 48 nodes in the longitudinal direction. Many such lattices (10–30) cover the globe in a stacked fashion to give a three-dimensional coverage of the atmosphere.

the U.S. National Meteorological Center (NMC), recently renamed the National Centers for Environmental Prediction (NCEP), and are also valid for the global prediction model at the European Centre for Medium Range Weather Forecasts (ECMWF). Note that distances between nodes can be a quarter of those quoted above for models of less than global extent used for short-range (0- to 3-day) forecasting.

From the values of the predicted quantities and the estimates of their spatial derivatives, all the terms in equations (1.1a), (1.1b), and (1.2–1.5) can be evaluated to determine the (local) time derivative of the forecasted quantities at the grid nodes. These (approximate) time derivatives are then used in a finite difference method to determine the prognostic quantities a short time in advance, typically about 15 minutes. Since such a short-range forecast is not of general use, the process is continued using the new values of the predicted quantities at the nodes to make a second forecast of 15 minutes' duration, then a third forecast, and so on until a potentially useful (12-hour to 10-day) forecast is arrived at.

The dramatic increase in computing power over the past 40 years has greatly influenced the accuracy of numerical predictions. This progress is illustrated in Figure 1.2 (top), which depicts the skill of 36-hour forecasts as a function of time since the inception of operational numerical weather prediction though 1986. Figure 1.2 (bottom) shows a recent skill record during winter for lead times ranging from 0 to 10 days. A significant reason for this improvement is the fact that with faster computers and larger storage capacity, models can be integrated with much finer mesh spacing than was previously possible. Thus major improvements in forecast skill mirror the major advances in computing technology.

3. Finite predictability

Despite this impressive progress in increasing skill, computer-produced weather forecasts are far from perfect. Imperfections are partially due to the fact that even with today's supercomputer technology, the distance between nodes is not sufficiently small to resolve (i.e., capture) the scale of phenomena responsible for thunderstorms and other weather features. Figure 1.2 represents a scientist's bias in that it depicts the improvement in forecast skill of upper-level flow patterns, approximately 5 km above the surface, which are associated with the high- and low-pressure patterns the media weather forecasters often show and which are resolved by current forecast models. Precipitation events associated with such phenomena are oftentimes one or two orders of magnitude smaller in horizontal extent, being structurally linked to the warm and cold fronts. Yet, for most people, precipitation is the single most important discriminator between a correct and incorrect forecast. Thus, the improvement in forecasts of surface weather, while still substantial, is not necessarily as impressive as Figure 1.2 implies. As will be explained below, both in the forecast equations and in the actual forecast issuance, a statistical procedure is used to incorporate the effects of phenomena too small in spatial scale to be resolved by the computer representation of the forecast equations (as well as to remove any systematic bias of the numerical model).

Hidden within the terms representing sources and sinks of heat, moisture, and momentum are representations of physical processes too small in scale and sometimes too complex to be completely included in a numerical forecast model. As mentioned previously,

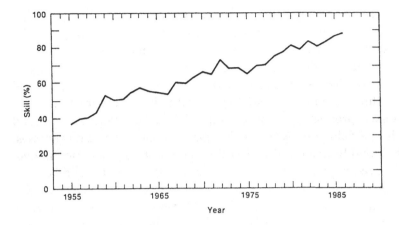

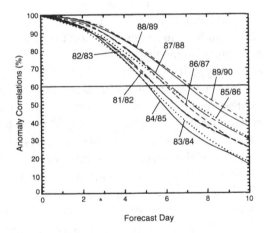

Figure 1.2. Record of forecast skill of 500 mb height over North America: (top) annual average skill of 36-hour forecasts (i.e., forecasts of 36-hour lead time) (Courtesy of U.S. National Meteorological Center); (bottom) recent skill for winter as a function of lead time (shown is the anomaly correlation, with separate curve for each winter season; horizontal line represents limit for synoptically useful forecasts). (From Kalnay, Kanamitsu, and Baker, 1990).

convective storms and their associated precipitation are an example of one such process. Other examples include clouds and their solar and infrared radiative interactions; turbulent exchanges of heat, moisture, and momentum resulting from the interaction of the atmosphere and earth's surface; and the momentum sink associated with gravity (buoyancy) waves induced by the small-scale topography of the earth. All these processes are primarily or exclusively active on spatial scales too small to be resolved in present-day models of the atmosphere. They are called "subgrid scale," for they represent processes occurring on scales that would fit within a volume surrounding a grid point in Figure 1.1.

If such processes influenced only the motion, temperature, and moisture fields on spatial scales characteristic of these processes, they could be neglected within the context of the prediction model. (The prediction of a thunderstorm would be statistically related to the prediction of the larger-scale model but would have no influence on the larger-scale model's prediction.) This is not the case, however. Thunderstorms, turbulent motions in the lowest kilometer of the atmosphere, and cloud radiation interactions all influence the larger scales through their averaged effects (transports of moisture, heat, and momentum for storms and turbulent gusts and mean correlations for radiation and clouds). Thus the statistical relationships must be incorporated in the physical model and used continuously to produce a large-scale forecast. This statistical empirical relation is called a "parameterization." The "art" of numerical weather prediction, as opposed to the science, is intimately tied to the parameterization of subgrid scale processes.

The second reason for imperfect weather forecasts is related to a fundamental aspect of the governing equations: their nonlinearity. The sole reason that thunderstorms and turbulent subgrid scale motions can influence their large-scale environments is that the prediction equations written out in a Eulerian frame contain product terms — such as the vertical transport of temperature, wT, and moisture, wq, by the atmosphere. These product terms have large-scale influence, even if individually each variable has only small-scale structure. This, in turn, necessitates the parameterization noted above, and leads to yet another difficulty: any small-scale errors, for example, on the scale of two or three grid volumes, will cause errors in the terms used to forecast all larger scales of the forecast model — even the global scale. This would not be disastrous to the large-scale forecast were it not for the fact that the equations governing the atmosphere are subject to a now commonly recognized, inherent difficulty: Their predictions are extremely sensitive to small changes in the predicted variables. This is the hallmark of a chaotic system with limited predictability.

The recognition of the unstable nature of the equations governing atmospheric evolution can be traced back to the ideas set forth in studies by P. D. Thompson and E. N. Lorenz in the late 1950s and early 1960s. Thompson's work (Thompson, 1957) was motivated by his early operational experience with numerical weather predictions, while Lorenz's efforts were related to his research in extended-range (monthly) prediction. Lorenz distilled the essence of the predictability problem in a seminal paper (Lorenz, 1963), in which he demonstrated that a system of only three nonlinear differential equations can have solutions that are sensitively dependent on their initial conditions. That a simple system of such low dimensionality (note that a typical computational model of the atmosphere requires the solution of several million such equations) could exhibit this behavior came as a surprise not only to the meteorological community, but also to the mathematics and mathematical physics communities. Although these latter groups began to recognize the ubiquity of this type of chaotic behavior a decade after Lorenz's publication, this system remains one of the prototypical examples of a deterministic system with limited predictability.

Other researchers in the field have followed Thompson's lead and applied the tools of the statistical theory of turbulence to the problem of predictability. The studies of D. K. Lilly, C. E. Leith, and R. H. Kraichnan in the early 1970s elucidated the nature of the loss of predictability inherent in the necessity of limiting the range of scales explicitly predicted and of parameterizing the subgrid scales (Kraichnan, 1970; Leith, 1971; Lilly, 1972). Such studies helped place upper bounds on the time interval over which weather forecasts have any skill. These early estimates suggested a two-week limit on forecast range, while more recent, more conservative estimates (cf. Lorenz, 1982; or Tribbia and Baumhefner, 1988) suggest about 10 days. Lorenz's 1982 work shows a comparison between actual and theoretically attainable forecast skill (Figure 1.3).

The output of a numerical forecast model predicts directly the following meteorological variables: three-dimensional variation in pressure, temperature, horizontal winds, and relative humidity; two-dimensional fields of surface pressure, temperature, and precipitation. These fields are available in time increments of approximately 20 minutes and have been listed in decreasing order

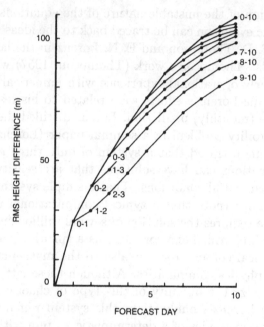

Figure 1.3. Global root-mean-square 500 mb height differences $E(j, k)$, in meters, between j-day and k-day forecasts made by the European Centre for Medium Range Weather Forecasts operational model for the same day, for $j < k$, plotted against k. A j-day forecast is one with a lead time of j days. Values of (j, k) are shown beside some of the points (e.g., "0–1" indicates $j = 0$ days and $k = 1$ days). Uppermost curve connects values of $E(0, k)$, $k = 1, 2, \ldots, 10$ days. Remaining curves connect values of $E(j, k)$ for constant $k - j$. (From Lorenz, 1982)

of accuracy at any given time within a forecast. Thus, pressure and temperature are the most accurate fields forecast, while moisture and precipitation are the most difficult for a numerical model to predict (Figure 1.4 shows trends in the accuracy of forecasts for temperature and precipitation).

4. The future

Figure 1.3 illustrates that the difference between what is being achieved and what can be achieved is not large, but its elimination would effectively double the range of useful predictive skill. The current research aimed at reducing this difference focuses on two goals: (i) to improve the numerical models so that they more accurately represent the physical processes in the atmosphere; and

(ii) to describe more accurately the initial state of the atmosphere as input to the computational iterations that result in the forecast. With respect to model improvements, it is clear that with increasing computer power, model resolution will increase, easing the need for parameterization of the current range of subgrid scales. Nevertheless, some processes are likely to remain beyond resolution for the foreseeable future (e.g., cloud formation, which occurs through droplet growth on the scale of microns), and subgrid scale parameterizations must be used and refined if models are to improve. However, increased resolution carries with it the need to observe the atmosphere on the spatial scale that the model resolves, so that accurate predictions of these small scales can be produced and evaluated. Thus, improving the accuracy of the initial input state of a computational model requires a more accurate specification of the currently resolved scales and an accurate specification of the smaller scales to be resolved in the future.

Fortunately, the resolution of scales of motion and the inclusion of such scales in a prediction model are linked by the manner in which atmospheric analyses used as initial states are produced. Because of the irregular manner in which the atmosphere is observed both temporally and spatially, interpolation in space to the grid nodes and in time to forecast initiation times is needed. Sophisticated statistical interpolation methods are currently used to interpolate in space, while the forecast model itself is used to interpolate in time. Current research at NCEP and most other operational centers in the world is focusing on making a completely consistent four-dimensional analysis system using the principles of optimal control theory (cf. Daley, 1991). This focus makes the inclusion of smaller scales in the initial state a natural consequence of current and planned higher-resolution forecast models.

The accuracy of any initial input state is limited, however, by the quantity and accuracy of the observations used in the interpolation method. In order to obtain more accurate large- and small-scale analyses, more detailed information is necessary. Currently, the backbone of the analysis system used for initial state construction is the land-based, upper-air radiosonde network developed as an outgrowth of World War II. Over the oceans, weather satellites afford coverage, but primarily they observe infrared radiation indicative of atmospheric temperature and humidity, requiring winds to be estimated from cloud tracks in the visible satellite channels.

An estimate of the accuracy of the current analysis system on the large scales included in present-day models can be obtained by extrapolating the forecast error curve in Figure 1.3 to day 0. New space-based and land-based sensors, such as lidar and radar wind profilers, are currently being developed and incorporated into the observing network. Wind, temperature, and moisture fields with high resolution in both space and time are needed to provide accurate small-scale and more accurate large-scale information to the analysis system used in numerical weather prediction.

Last, it has become increasingly clear that even if detailed accuracy is unachievable because of the inevitable growth of small initial errors, some useful information may be gained from numerical weather forecasts if the forecast system recognizes the statistical nature of the problem. This is already being done to a certain degree in precipitation forecasts (as well as other variables) where, as noted above, convective precipitation is a result of a parameterized physical process. A model forecast of the occurrence of convective precipitation should actually be regarded as indicative of a higher likelihood than normal of this type of precipitation within the grid volume. To optimize skill in forecasting precipitation locally, operational forecast centers have developed and utilized statistical relationships between the output of numerical models and forecasts of precipitation and other weather variables for individual forecast locales. Such use of model output statistics (MOS) and other statistical approaches is discussed at length in Wilks (1995, chap. 6); also Chapters 2 through 5 in the present volume include consideration of forecasts produced by this technique.

For the purpose of completeness, here it is sufficient to note that the various approximations made in producing a numerical forecast for a given locale, both physical and computational in origin, can lead to systematic errors in forecasts of grid cell–averaged weather. This error can be corrected, in the mean, through the use of MOS, which interprets the forecast of grid cell–averaged weather in the cell containing the specific locale of interest — not as the weather to be expected, but as predictors of the weather to be input into a statistical forecast model. This statistical model can be used to correct the systematic bias of the numerical forecast model and to downscale, or refine, the forecast to the locale of interest. Additionally, quantities that can be forecast sensibly only in terms of probabilities are relatively easily handled by specifying the probability of occurrence as the predictand in the MOS formulation. Improvements in the accuracy of MOS predictions in recent years are shown for maximum/minimum temperature and probability of precipitation (Figure 1.4).

Another quantitative result from the statistical turbulence studies of the early 1970s was that smaller scales lose predictive skill first and errors progress up scale, until the largest scales are finally contaminated with error and the forecast is totally devoid of skill (Lorenz, 1969). Thus precipitation, which can occur on very small scales of the order of a kilometer, can be accurately predicted only for a short period of time on the order of a few hours, while a large-scale low-pressure disturbance in midlatitudes can be accurately forecast for up to a week or more. In both cases, a completely accurate forecast requires the exact determination of the location and timing of the event. Nevertheless, forecasts with less than perfect skill, as reflected by probability forecasts produced using model predictions, should still be societally valuable.

Weather services are currently researching the possibility of meeting these needs by experimentally producing ensembles or multiple realizations of numerical predictions (Toth and Kalnay, 1993). This approach allows a forecaster to consider the relative likelihood of a particular weather event's occurrence by examining its frequency of occurrence in the forecast ensemble.

The theoretical roots of such methods were first discussed by E. S. Epstein and C. E. Leith in the early 1970s (Epstein, 1969; Leith, 1974). After two decades of gradual improvement in computer power, the advantages of increasing the resolution of forecast models can now be weighed against the advantages of probabilistic ensemble predictions. As stated above, this trade-off is a function of the spatial and temporal scale of the phenomenon to be predicted. For convective precipitation, ensemble forecasts should be of use in determining rainfall risks in forecasts of one day or longer, while for global-scale weather, ensembles should be of use for forecasts longer than one week. Because the predictability error growth studies (noted above) have shown that small-scale features inevitably are the first to lose skill in a forecast, high-resolution information is of little use beyond these phenomena-dependent time ranges. Thus, computational effort and expense is more wisely

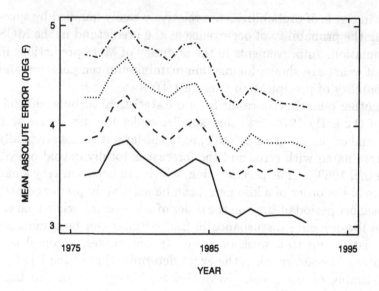

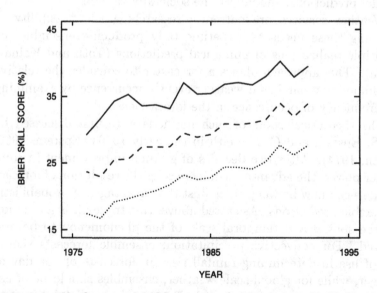

Figure 1.4. Verification of MOS forecasts for approximately 90 National Weather Service reporting stations (solid line represents 12- to 24-hour forecast, dashed line represents 24- to 36-hour forecast, dotted line represents 36- to 48-hour forecast, and dot-dashed line represents 48- to 60-hour forecast: (top) mean absolute error for maximum/minimum temperature; (bottom) Brier skill score (see Chapter 2 of this volume) for probability of precipitation. (From Vislocky and Fritsch, 1995)

utilized in the production of multiple forecasts. Such methods allow the numerical model to determine which forecast features are robust in the presence of the noise of predictability error. They also have obvious applicability in forecasting the atmospheric response to anomalous atmospheric forcing, for example, the equatorial sea surface temperature anomalies associated with El Niño, which form the basis for current operational monthly and seasonal outlooks.

Acknowledgments

The author wishes to thank the late P. D. Thompson, whose accounts of the historical developments in the field of numerical weather prediction form much of the content of this chapter. Of course, any misstatements of fact are the author's responsibility.

References

Ashford, O.M. (1985). *Prophet – or Professor? The Life and Work of Lewis Fry Richardson*. Bristol: Adam Hilger.

Bjerknes, V. (1904). The problem of weather forecasting considered from the point of view of mathematics and mechanics. *Meteorologische Zeitschrift*, **21**, 1–7.

Daley, R. (1991). *Atmospheric Data Analysis*. Cambridge: Cambridge University Press.

Epstein, E.S. (1969). Stochastic dynamic prediction. *Tellus*, **11**, 739–759.

Kalnay, E., Kanamitsu, M. & Baker, W.E. (1990). Global numerical weather prediction at the National Meteorological Center. *Bulletin of the American Meteorological Society*, **71**, 1410–1428.

Kraichnan, R.H. (1970). Instability of fully developed turbulence. *Physics of Fluids*, **13**, 569–575.

Leith, C.E. (1971). Atmospheric predictability and two dimensional turbulence. *Journal of the Atmospheric Sciences*, **28**, 145–161.

Leith, C.E. (1974). Theoretical skill of Monte-Carlo forecasts. *Monthly Weather Review*, **102**, 409–418.

Lilly, D.K. (1972). Numerical simulation studies of two-dimensional turbulence II. *Geophysical Fluid Dynamics*, **4**, 1–28.

Lorenz, E.N. (1963). Deterministic nonperiodic flow. *Journal of the Atmospheric Sciences*, **20**, 130–141.

Lorenz, E.N. (1969). The predictability of a flow which possesses many scales of motion. *Tellus*, **21**, 289–307.

Lorenz, E.N. (1982). Atmospheric predictability experiments with a large numerical model. *Tellus*, **34**, 505–513.

Lynch, P. (1994). Richardson's marvellous forecast. *Proceedings, International Symposium on the Life Cycles of Extratropical Cyclones*, 38–48. Bergen, Norway.

Petterssen, S. (1969). *Introduction to Meteorology* (third edition). New York: McGraw-Hill.

Richardson, L.F. (1922). *Weather Prediction by Numerical Process*. London: Cambridge University Press.

Thompson, P.D. (1957). Uncertainty of initial state as a factor in the predictability of large scale atmospheric flow pattern. *Tellus*, **9**, 275–295.

Toth, Z. & Kalnay, E. (1993). Ensemble forecasting at NMC: the generation of perturbations. *Bulletin of the American Meteorological Society*, **74**, 2317–2330.

Tribbia, J.J. & Baumhefner, D.P. (1988). Estimates of predictability of low-frequency variability with a spectral general circulation model. *Journal of the Atmospheric Sciences*, **45**, 2306–2317.

Vislocky, R.L. & Fritsch, J.M. (1995). Improved model output statistics forecasts through model consensus. *Bulletin of the American Meteorological Society*, **76**, 1157–1164.

Wilks, D.S. (1995). *Statistical Methods in the Atmospheric Sciences*. San Diego: Academic Press.

2

Forecast verification

ALLAN H. MURPHY

1. Introduction

Forecast verification has had a long, colorful, and occasionally controversial history. The first substantial developments in verification methods were occasioned by the publication of a paper by J.P. Finley (Finley, 1884), in which he summarized some results of an experimental tornado forecasting program. As a measure of forecasting success, Finley calculated the sum of the percentages of correct forecasts of tornadoes and no tornadoes (multiplied by 100) and reported a percentage correct value of 96.6%. Finley's paper attracted the attention of several individuals (e.g., Doolittle, 1885; Gilbert, 1884; Peirce, 1884), and (*inter alia*) it was pointed out that always forecasting no tornadoes would have led to a percentage correct value of 98.2%! These and other individuals offered various criticisms of Finley's method of verification and proposed alternative measures of overall forecasting performance, most of which are still in use today. This signal event in the history of forecast verification — dubbed the *Finley affair* — is described in detail in a recent paper (Murphy, 1996).

The 100+ years since the Finley affair have witnessed the development of a relatively wide variety of methods and measures tailored to particular verification problems. Moreover, considerable attention has been devoted in recent years to studies related to the properties of various methods/measures. Space limitations preclude any attempt to review these developments here. Readers interested in the history of forecast verification, or the numerous methods and measures formulated since 1884, are referred to Brier and Allen (1951), Meglis (1960), Murphy and Daan (1985), Stanski, Wilson, and Burrows (1989), and Wilks (1995), as well as the references cited therein. Some of the methods developed during this period will be introduced in subsequent sections of the chapter.

Forecast verification is usually defined as the process of assessing the degree of correspondence between forecasts and observations (e.g., Murphy and Daan, 1985). In practice, the verification process has generally consisted of (i) calculating quantitative measures of one or two aspects of forecasting performance such as bias, accuracy, or skill (these and other terms are defined in Section 2) and (ii) drawing conclusions regarding absolute and/or relative performance on the basis of numerical values of these measures. Thus, traditional methods can be said to constitute a *measures-oriented (MO) approach* to the verification process.

In this chapter, forecast verification is defined as the process of assessing the quality of forecasts, where forecast quality is understood to consist of the totality of statistical characteristics embodied in the joint distribution of forecasts and observations (Murphy and Winkler, 1987). This definition leads to a *distributions-oriented (DO) approach* to the verification process, an approach that appears to offer clear theoretical and practical advantages over the traditional MO approach. For a recent case study contrasting the MO and DO approaches to forecast verification, see Brooks and Doswell (1996). The concepts, methods, and measures described here are based on this DO approach to forecast verification.

An important advantage of the DO approach is that it imposes some structure on verification problems themselves as well as on the body of verification methodology. With regard to the latter, it provides insight into the relationships among verification methods and measures, and it creates a sound basis for developing and/or choosing particular methods/measures in specific applications. From a practical point of view, this approach places particular emphasis on the assessment of various basic aspects of forecast quality, thereby providing the producers of forecasts (e.g., modelers, forecasters) with information that they need to guide their efforts to improve forecasting performance.

This chapter focuses on concepts and methods related to the verification of forecasts of weather variables. The body of verification methods described here is applicable to all types of forecasts — for example, probabilistic *and* nonprobabilistic forecasts — as long as they are evaluated on a location-by-location basis (or the forecasts from different locations are pooled to create a single verification data sample). The framework upon which these methods are based is consistent with the multidimensional structure of verification problems and the multifaceted nature of forecast quality. These considerations are important in evaluating and comparing forecasting systems to avoid reaching inappropriate conclusions regarding absolute or relative forecasting performance. A distinction is made here between the problems of *absolute verification* (i.e., evaluating the forecasts produced by a single forecasting system) and *comparative verification* (i.e., evaluating and comparing the forecasts produced by two or more forecasting systems).

Section 2 describes the basic elements of the conceptual and methodological framework upon which the suite of DO verification methods to be presented here are based. These elements include the verification data sample, the joint distribution of forecasts and observations, the factorization of this distribution into conditional and marginal distributions, the complexity and dimensionality of verification problems, forecast quality and its various aspects, the nature of verification measures, and criteria for screening verification measures. General DO methods appropriate to the problems of absolute and comparative verification are described in some detail in Sections 3 and 4, respectively. To illustrate the use and interpretation of the basic suite of DO verification methods, Sections 3 and 4 include applications of these methods to samples of nonprobabilistic temperature and precipitation probability forecasts. Other verification problems and methods are briefly considered — and references to relevant works are provided — in Section 5. Section 6 contains a short discussion of the relationship between forecast quality and forecast value. Section 7 includes a summary and some concluding remarks.

2. Conceptual and methodological framework

2.1. *Forecasts, observations, and verification data samples*

The verification problems of interest here involve the evaluation and comparison of forecasting systems (forecasting methods, forecasting models, forecasters). These systems are denoted by F, G, ... Generic forecasts produced by systems F, G, ..., are denoted by f, g, ..., respectively, and these forecasts represent descriptions of future weather conditions.

To evaluate and compare forecasting systems, observations that describe the actual weather conditions (i.e., the weather conditions that occur at or during the valid time of the forecasts) must be available. Here, the observations are assumed to represent sets of weather situations characterized by variables denoted by X, Y, ... Generic values of these observations are denoted by x, y, ..., respectively.

The underlying weather variables, to which both the forecasts and observations refer, possess various basic characteristics. Some of these variables are continuous quantities (e.g., temperature, precipitation amount) and others are discrete quantities (e.g., cloud amount, precipitation type). However, all weather variables are generally treated as discrete — and defined in terms of a finite set of categories (of values) or events — for the purposes of prediction and evaluation. This practice is followed here, and the body of verification methods described in Sections 3 and 4 can be applied to forecasts of all (discrete) weather variables.

The forecasts themselves can be expressed in a probabilistic or nonprobabilistic format. Probabilistic forecasts indicate the likelihood of occurrence of the various possible categories or events that define the underlying variable, whereas nonprobabilistic forecasts specify that a particular category or event will occur (with complete certainty). The observations are assumed to be expressed in a nonprobabilistic format.

In the case of nonprobabilistic forecasts, the sets of possible forecasts and observations are generally identical. However, in the case of probabilistic forecasts, the set of distinct forecasts is usually larger (often considerably larger) than the set of observations. The implications of this difference for the verification process are discussed in Section 2.4.

Traditionally, the *verification data sample* associated with forecasting system F and forecasting situations X is viewed as a sample of n matched pairs of forecasts and observations $\{(f_i, x_i), \ i = 1, \ldots, n\}$, where n is the sample size. In this chapter the verification data sample is arranged instead by distinct pairs of forecasts and observations. For example, a sample of temperature forecasts and observations might contain some occasions on which the forecast temperature is 60°F and the observed temperature is 58°F, as well as other occasions on which the forecast temperature is 60°F and the observed temperature is 63°F. These two subsamples correspond to distinct combinations of forecast and observed temperatures. The phrase "all (f, x)-pairs" is used to denote this arrangement of the verification data sample.

It is convenient here to introduce some basic notation to describe the verification data sample. Let n^f and n^x denote the number of distinct forecasts and observations, respectively. Moreover, let n_{ij} denote the joint frequency of forecast f_i and observation x_j ($i = 1, \ldots, n^f$; $j = 1, \ldots, n^x$). Then the n^f-by-n^x matrix $\underline{N} = (n_{ij})$ contains the joint frequencies of all (f, x)-pairs in the verification data sample. The marginal frequencies of forecast f_i and observation x_j are denoted by $n_i^f = \sum_j n_{ij}$ and $n_j^x = \sum_i n_{ij}$, respectively ($\sum_i n_i^f = \sum_j n_j^x = \sum_i \sum_j n_{ij} = n$).

2.2. Joint distribution of forecasts and observations

The joint distribution of forecasts and observations, denoted here by $p(f, x)$, constitutes the basic framework for the DO approach to forecast verification (Murphy and Winkler, 1987). This joint distribution contains information about the forecasts, about the observations, and about the relationship between the forecasts and observations. Under the assumptions that the bivariate time series of forecasts and observations is (i) serially independent and (ii) stationary in a statistical sense, the distribution $p(f, x)$ contains all the relevant information in the verification data sample. *Forecast quality*, as defined here, is the totality of the statistical characteristics of the forecasts, the observations, and their relationship embodied in this distribution.

Let p_{ij} denote the joint relative frequency of forecast f_i and observation x_j in the verification data sample ($i = 1, \ldots, n^f$; $j = 1, \ldots, n^x$). It follows that $p_{ij} = n_{ij}/n$. Moreover, the n^f-by-n^x matrix $\underline{P} = (p_{ij})$ contains the joint relative frequencies of all (f, x)-pairs. These joint relative frequencies are treated here as joint probabilities (i.e., the estimation problem is ignored). Thus, $p_{ij} = \Pr(f = f_i, x = x_j)$, and the matrix $\underline{P}$ contains the joint probabilities of all (f, x)-pairs. These probabilities can be viewed as parameters of a primitive model of the joint distribution $p(f, x)$ (see Section 2.5).

As noted above, the joint distribution $p(f, x)$ plays a fundamental role in the verification process. In essence, forecast verification

consists of describing and summarizing the statistical characteristics of this distribution. This process is facilitated by factoring
the joint distribution into conditional and marginal distributions.

2.3. Factorizations of joint distribution: conditional and marginal distributions

The information relevant to forecast quality contained in the joint
distribution $p(f, x)$ becomes more accessible when this distribution
is factored into conditional and marginal distributions. Two such
factorizations can be defined:

$$p(f, x) = q(x|f)s(f) \qquad (2.1)$$

and

$$p(f, x) = r(f|x)t(x) \qquad (2.2)$$

(Murphy and Winkler, 1987). The expression in equation (2.1),
the *calibration-refinement (CR) factorization* of $p(f, x)$, involves
the conditional distributions of the observations given the fore-
casts, $q(x|f)$, and the marginal distribution of the forecasts, $s(f)$.
A conditional distribution $q(x|f)$ can be defined for each of the n^f
possible forecasts, with $q(x_j|f_i) = q_{ij} = \Pr(x = x_j|f = f_i)$. These
conditional distributions can be depicted in the form of an n^f-by-
n^x matrix $\underline{Q} = (q_{ij})$, where the elements of the ith row are the
components of the conditional distribution $q(x|f_i)$. The marginal
distribution $s(f)$ specifies the unconditional probabilities of the n^f
forecasts. That is, $s(f_i) = s_i = \Pr(f = f_i) [= \Sigma_j p_{ij} (j = 1, \ldots, n^x)]$.
This distribution can be depicted in the form of a n^f-by-1 vector
$\underline{s} = (s_i)$.

The expression in equation (2.2), the *likelihood-base rate (LBR)
factorization* of $p(f, x)$, involves the conditional distributions of
the forecasts given the observations, $r(f|x)$, and the marginal
distribution of the observations, $t(x)$. A conditional distribution
$r(f|x)$ can be defined for each of the n^x possible observations, with
$r(f_i|x_j) = r_{ij} = \Pr(f = f_i|x = x_j)$. These conditional distributions
can be depicted in the form of an n^f-by-n^x matrix $\underline{R} = (r_{ij})$, where
the elements of the jth column are the components of the distribu-
tion $r(f|x_j)$. The marginal distribution $t(x)$ specifies the uncondi-
tional probabilities of the n^x observations. That is, $t(x_j) = t_j =$

$\Pr(x = x_j)$ $[= \Sigma_i p_{ij}(i = 1,\ldots,n^f)]$. This distribution can be depicted in the form of a 1-by-n^x vector $\underline{t} = (t_j)$.

Since the joint distribution $p(f,x)$ can be reconstructed from the components on the right-hand side of either equation (2.1) or equation (2.2), forecast verification based on either factorization is equivalent to forecast verification based on $p(f,x)$ itself. In fact, DO verification can be conducted within frameworks based on $p(f,x)$, $q(x|f)$ and $s(f)$, or $r(f|x)$ and $t(x)$. However, since the three DO frameworks provide insights into different aspects of forecast quality (see Section 2.6), it is more appropriate to view these frameworks as complementary rather than alternative approaches to verification problems. Of course, the fundamental equivalence of the frameworks embodied in equations (2.1) and (2.2) implies that some relationships must exist among these various aspects of quality (see Murphy and Winkler, 1987).

2.4. Verification problems and their complexity

Verification problems are of two basic types: (i) *absolute verification* and (ii) *comparative verification* (Murphy, 1991). Absolute verification problems are concerned with the evaluation of forecasts produced by individual forecasting systems. In the case of a forecasting system F and forecasting situations described by X, absolute forecast verification is based on the joint distribution $p(f,x)$ and the conditional and marginal distributions associated with the factorizations of $p(f,x)$. Absolute verification also includes the comparison of the forecasts of interest with naive forecasts derived (solely) from the marginal distribution of observations $t(x)$ (e.g., forecasts based on sample climatology or possibly persistence).

Comparative verification problems involve the evaluation and comparison of two or more forecasting systems. If two forecasting systems, denoted by F and G, formulate forecasts for the same set of forecasting situations (i.e., a common set of situations described by X), then comparative verification consists of evaluating and comparing $p_F(f,x)$ and $p_G(g,x)$ and the components of their respective factorizations. On the other hand, if F's and G's forecasts relate to different forecasting situations (i.e., different sets of situations X and Y), then comparative verification consists of evaluating and comparing $p_{F,X}(f,x)$ and $p_{G,Y}(g,y)$, as well as the components of their respective factorizations. These two types of comparative verification are referred to as *matched comparative verification* and *unmatched comparative verification.*

The complexity of verification problems can be characterized by the number of distinct sets of quantities (i.e., forecasts and observations) associated with the underlying joint distribution(s). According to this definition, absolute verification problems involve two sets of quantities and comparative verification problems involve three or four (or more) sets of quantities. Thus, the latter are more complex than the former (Murphy, 1991). Moreover, within the context of comparative verification, unmatched comparative verification is more complex than matched comparative verification. The treatment of comparative verification in this chapter is restricted to the case of matched comparative verification.

Reductions in the complexity of verification problems can be achieved by decreasing the number of these basic quantities. For example, a considerable reduction in the complexity of absolute verification problems can be accomplished by assuming that attention can be focused on the (univariate) distribution of forecast errors, $u(e)$ $[= u(f - x)]$, rather than the joint distribution of forecasts and observations, $p(f, x)$. Whether or not such a strong assumption is warranted depends on the verification problem at hand (including the nature of the underlying variable). In any case, a large amount of potentially useful information regarding various aspects of forecast quality becomes inaccessible when absolute verification is based on $u(e)$ rather than on $p(f, x)$.

2.5. Models of basic distributions: dimensionality of verification problems

The dimensionality (d) of an absolute verification problem relates to the number of probabilities (or parameters) that must be determined in order to reconstruct the basic joint distribution $p(f, x)$ (Murphy, 1991). In situations in which this distribution is described by the elements of the matrix of joint probabilities, $\underline{P} = (p_{ij})$,

$$d = n^f n^x - 1, \tag{2.3}$$

since the sum over all elements in $\underline{P}$ must equal unity. Equivalent definitions of d can be formulated in terms of the elements of the matrix Q or R and vector s or t describing the conditional and marginal distributions, respectively.

Thus, an absolute verification problem involving nonprobabilistic forecasts for a two-category (or dichotomous) variable is a three-dimensional problem $(d = 2 \times 2 - 1)$. That is, it is necessary to determine three joint probabilities (or, for example, two conditional probabilities and one marginal probability) to describe forecast quality completely in this situation. On the other hand, an absolute verification problem involving probabilistic forecasts for a dichotomous variable, in which 11 equally spaced probability values are used, is a 21-dimensional problem $(d = 11 \times 2 - 1)$. Twenty-one joint probabilities (or, for example, 11 conditional probabilities and 10 marginal probabilities) must be determined in order to describe forecast quality completely in this situation. The corresponding comparative verification problems possess considerably higher (approximately two or more times higher) dimensionality, since they involve two or more joint distributions.

When viewed from this perspective, most verification problems possess relatively high dimensionality. In general, the higher the dimensionality of a verification problem, the more quantities (i.e., joint, conditional, and/or marginal probabilities, or numerical measures) must be determined in order to provide a complete — or even an adequate — description of forecast quality. Clearly, a single overall measure of forecasting performance, such as the mean square error, cannot provide a complete — or even a very insightful or useful — description of forecast quality in most problems. Considerations related to the dimensionality of verification problems have generally been ignored in the traditional MO approach to forecast evaluation. To provide a reasonably complete description of forecast quality, the dimensionality of verification problems must be respected.

The description of the distribution $p(f, x)$ in terms of joint probabilities represents a model — albeit a primitive model — of this distribution (equivalent statements could be made with respect to the probabilities that constitute the conditional and marginal distributions). This primitive model is consistent with the usual approach to forecast verification, in the sense that the verification process is based on the empirical frequencies (or relative frequencies) derived from the verification data sample. Since these primitive models generally possess many parameters (in this case, joint, conditional, and/or marginal probabilities), the verification process suffers from the "curse of dimensionality."

Although the verification methods to be described in this chapter are based — explicitly or implicitly — on this primitive model, it may be useful to consider briefly the possibility of reducing the dimensionality of verification problems by modeling the basic distributions with parametric statistical models. If acceptable fits to the basic (empirical) distributions could be obtained, then such models would provide parsimonious descriptions of the corresponding verification data samples. Forecast quality could then be characterized in terms of a relatively few model parameters, thereby substantially reducing the dimensionality of these verification problems. Moreover, characterizing forecast quality in terms of the parameters of one or more statistical models should reduce the undesirable effects of sampling variability on assessments of forecast quality (and its aspects). Currently, these effects are generally ignored, even though it is recognized that another verification data sample acquired under similar conditions will yield at least somewhat different results. Reducing the effects of sampling variability would lead to more credible estimates of the various aspects of forecast quality. Finally, the availability of models of forecast quality should facilitate studies of the relationship between forecast quality and forecast value (e.g., Katz and Murphy, 1990; Katz, Murphy, and Winkler, 1982; see also Chapter 6 in this volume).

Most studies in which statistical models have been used to describe forecast quality are of relatively recent vintage. Moreover, many of these models were identified in the context of forecast-value studies rather than in the context of forecast verification problems. For example, Katz et al. (1982) used a bivariate Gaussian model to characterize the relationship between daily minimum temperature forecasts expressed in a nonprobabilistic format and the corresponding observations in the context of a fruit-frost decision-making problem. In this case a potential verification problem of relatively high dimensionality was reduced to no more than 5 dimensions ($d \leq 5$), represented by two means, two variances, and one covariance (or correlation coefficient).

More recently, Krzysztofowicz and Long (1991b) investigated the use of the sufficiency relation (see Section 4.1) as a means of comparing the performance of objective and subjective precipitation probability forecasts. To simplify the process of evaluation, they used beta distributions to fit the conditional distributions (or likelihoods) $r(f|x = 1)$ and $r(f|x = 0)$ for both types of forecasts. This approach reduced a comparative verification problem of approximately 40 dimensions to a problem of 8 dimensions $(d = 8)$, represented by 4 parameters associated with the conditional distributions for each type of forecast.

As noted previously, it may not always be possible to identify mathematically convenient statistical models that fit the relevant joint, conditional, and/or marginal distributions in a satisfactory manner. For example, Clemen and Winkler (1987) used a Gaussian log-odds model to fit the likelihood functions — that is, $r(f|x = 0)$ and $r(f|x = 1)$ — for samples of precipitation probability forecasts and the corresponding observations in a calibration and combining study. They found that these models tended to yield distributions that were appreciably more skewed than the empirical distributions.

In summary, the use of statistical models offers a means of describing forecast quality in a parsimonious manner, thereby simplifying many verification problems. This approach — a substantial departure from the traditional approach (involving the use of empirical relative frequencies) adopted here — clearly warrants further investigation. Some results of a recent effort to use statistical models to reduce the dimensionality of verification problems and the effects of sampling variability, in the context of precipitation probability forecasting, are reported by Murphy and Wilks (1996).

2.6. Forecast quality and its aspects

As noted in Section 2.2, forecast quality is defined as the totality of the statistical characteristics of the forecasts, the observations, and their relationship embodied in the joint distribution $p(f, x)$. Moreover, specific aspects of quality can be related to $p(f, x)$ or to the conditional and marginal distributions associated with its factorizations. These aspects of quality are identified and defined in Table 2.1. This table also indicates the basic distributions associated with each aspect of quality.

Bias (also referred to as systematic or unconditional bias) relates to the degree of correspondence between the average forecast

Table 2.1. Names and definitions of aspects of quality, including basic distribution(s) related to each aspect

Name	Definition	Basic distribution(s)
Bias	Degree to which μ_f corresponds to μ_x	$s(f)$ and $t(x)$
Association	Overall strength of linear relationship between f and x	$p(f,x)$
Accuracy	Average degree of correspondence between f and x	$p(f,x)$
Skill (relative accuracy)	Accuracy of forecasts relative to accuracy of forecasts based on standard of reference (e.g., climatology)	$p(f,x)$
Type 1 conditional bias (reliability, calibration)	Degree of correspondence between $\mu_{x\|f}$ and f, averaged over all values of f	$q(x\|f)$ and $s(f)$
Resolution	Degree to which $\mu_{x\|f}$ differs from μ_x, averaged over all values of f	$q(x\|f)$ and $s(f)$
Sharpness (refinement)	Degree to which probability forecasts approach zero or one	$s(f)$
Type 2 conditional bias	Degree of correspondence between $\mu_{f\|x}$ and x, averaged over all values of x	$r(f\|x)$ and $t(x)$
Discrimination	Degree to which $\mu_{f\|x}$ differs from μ_f, averaged over all values of x	$r(f\|x)$ and $t(x)$
Uncertainty (variability)	Degree of variability in observations	$t(x)$

μ_f and the average observation μ_x. This aspect of quality is generally measured in terms of the difference between μ_f and μ_x. For example, if $\mu_f = 65.4°F$ and $\mu_x = 63.8°F$ for a sample of non-probabilistic temperature forecasts, then these forecasts exhibit a positive bias of 1.6°F.

Association refers to the strength of the linear relationship between the forecasts and observations. This aspect of quality is usually measured by the correlation coefficient ρ_{fx}. The square of ρ_{fx} represents the proportionate reduction in the variance of the observations when they are regressed on the forecasts.

Accuracy relates to the average degree of correspondence between individual forecasts and observations in the verification data sample. It is generally defined in terms of the joint distribution $p(f, x)$, but it also can be defined in terms of conditional and marginal distributions. In the context of forecast verification, common measures of accuracy include the mean square error or mean absolute error in the case of (essentially) continuous variables and the fraction (or percentage) of correct forecasts in the case of discrete variables.

Skill is usually defined as the accuracy of the forecasts of interest relative to the accuracy of forecasts produced by a naive forecasting system such as climatology or persistence. Skill scores, defined as measures of relative accuracy, are used to assess this aspect of quality. Positive (negative) skill scores indicate that the accuracy of the forecasts of interest is greater (less) than the accuracy of the forecasts produced by the standard of reference. In the case of operational temperature forecasts that possess a mean square error of 3.6 (°F)^2, when temperature forecasts based on persistence possess a mean square error of 5.4 (°F)^2, the skill of the forecasts is positive with a numerical skill score value of 0.333 [$= 1 - (3.6/5.4)$].

Reliability (calibration or type 1 conditional bias) relates to the degree of correspondence between the mean observation given a particular forecast, $\mu_{x|f}$, and the forecast f. Suppose that $\mu_{x|f} = 0.424$ for a subsample of precipitation probability forecasts for which $f = 0.40$. Clearly, this subsample of forecasts is not perfectly reliable (or conditionally unbiased). Forecasts that exhibit perfect correspondence between $\mu_{x|f}$ and f over all values of f are said to be completely reliable (or, equivalently, well-calibrated or conditionally unbiased overall). Forecasts that are completely reliable are necessarily unbiased (but the converse relationship does not hold).

Resolution relates to the difference between the mean observation given a particular forecast, $\mu_{x|f}$, and the overall unconditional mean observation μ_x. A verification data sample for which

$\mu_{x|f} = \mu_x$ for all f is completely lacking in resolution. Thus, larger differences between $\mu_{x|f}$ and μ_x are preferred to smaller differences. Resolution, as an aspect of forecast quality, is based on the concept that "different forecasts should be followed by different observations."

Sharpness (refinement) is an aspect of forecast quality that applies only to probabilistic forecasts. Such forecasts are perfectly sharp (refined) if only probabilities of zero and one are used in the forecasts. (Nonprobabilistic forecasts are always perfectly sharp.) On the other hand, constant forecasts of the climatological probability are completely lacking in sharpness. Since the degree of sharpness increases as more frequent use is made of relatively high and low probabilities, the variability of the forecasts is an indicator of sharpness. Forecasts involving probabilities of zero and one maximize this variability. Not surprisingly, the variance of the distribution of forecasts, σ_f^2, is frequently used as a one-dimensional measure of sharpness. When the forecasts of interest are completely reliable, sharpness and resolution become identical aspects of quality.

Type 2 conditional bias relates to the degree of correspondence between the mean forecast given a particular observation, $\mu_{f|x}$, and the observation x. Suppose that, in the case of precipitation probability forecasting systems A and B, $\mu_{f|x}(A) = 0.72$ and $\mu_{f|x}(B) = 0.68$ when $x = 1$ and $\mu_{f|x}(A) = 0.24$ and $\mu_{f|x}(B) = 0.32$ when $x = 0$. Then system A's forecasts are less conditionally biased in the type 2 sense than system B's forecasts. Forecasts that exhibit complete correspondence between $\mu_{f|x} = x$ for all x are said to be completely conditionally unbiased in this sense (probabilistic forecasts that satisfy this condition are necessarily perfect forecasts).

Discrimination relates to the difference between the mean forecast given a particular observation, $\mu_{f|x}$, and the overall mean forecast μ_f. In the hypothetical situation involving forecasting systems A and B considered above, A's forecasts exhibit greater discrimination than B's forecasts. When $\mu_{f|x} = \mu_f$ for all x the verification data sample is completely lacking in discrimination. Thus, larger differences between $\mu_{f|x}$ and μ_f are preferable to smaller differences.

Uncertainty relates to the variability of the observations (as primitive descriptors of the forecasting situations). The variance of the distribution of observations, σ_x^2, is sometimes used as a measure of uncertainty. It should be noted that this aspect of quality does not depend in any way on the forecasts. Thus, uncertainty is a characteristic of the observations, or forecasting situations, rather than a characteristic of the forecasts.

2.7. Measures of aspects of quality: verification measures

A *verification measure* is defined here as any (mathematical) function of the forecasts, the observations, or their relationship. Thus, the variance of the forecasts and the variance of the observations are verification measures, even though they are not concerned directly with the correspondence between forecasts and observations. Clearly, these two variances — and their relative magnitude — are of considerable interest in the context of many verification problems.

A *performance measure* is defined here as a verification measure that focuses on the correspondence between forecasts and observations, on either an individual or collective basis. The mean square error (a measure of accuracy), as well as the measure of bias defined as the difference between the mean forecast and the mean observation, represent examples of performance measures. Obviously, performance measures constitute a subset of the set of all verification measures.

A *scoring rule* is a performance measure that is defined for individual pairs of forecasts and observations. Thus, the mean square error is a scoring rule, whereas the measure of bias defined in the previous paragraph is not a scoring rule. Scoring rules represent a subset of the set of performance measures.

The set of all verification measures is essentially infinite. For the purposes of this chapter, two verification measures are considered to be equivalent if they are linearly related. This definition of equivalence is used in conjunction with the discussion of screening criteria for verification measures (see Section 2.8). Other definitions of equivalence may be used in other contexts.

As defined, verification measures in general — and performance measures in particular — may possess positive or negative orientations. In the case of performance measures, positive (negative) orientation implies that larger (smaller) scores are indicative of better performance (with regard to the aspect of quality of interest). Since a linear transformation of a performance measure leads to an equivalent measure, it is always possible to change the orientation of a performance measure from negative to positive (or vice versa).

It is sometimes desirable to establish a standard range of values (or scores) for a verification or performance measure. If the range of values of a measure are finite, then it can be transformed linearly in such a way as to produce any desired range of values. Ranges frequently considered desirable include [0, 1] with one (zero) representing the best (worst) possible score or [−1, +1] with plus (minus) one representing the best (worst) possible score. Some measures (e.g., skill scores; see Section 3.2) do not possess finite ranges of values; in this case, it is not possible to transform the measures in such a way as to obtain a desired finite range of values.

2.8. Criteria for screening verification measures

Four criteria that can be used to screen alternative verification measures are briefly described in this section. Each criterion relates to a desirable property or characteristic that a particular measure may or may not possess. Application of screening criteria is not intended to identify the single best measure (this goal is generally inappropriate and unattainable), but rather to eliminate from further consideration those measures that do not possess one or more of these desirable characteristics.

Sufficiency. The concept of sufficiency — and the sufficiency relation — will be considered in detail in conjunction with the discussion of comparative evaluation of forecasting systems in Section 4 (see Section 4.1). In brief, the sufficiency relation identifies the conditions under which one forecasting system can be judged to be unambiguously superior to another forecasting system (i.e., superior in terms of both quality and value). Under certain conditions, it may be possible to use this relation as a criterion for screening verification measures. Verification measures that are consistent with the sufficiency relation are generally preferred to measures that are not consistent with this relation.

For example, Krzysztofowicz (1992) has shown that in the case of nonprobabilistic forecasts for a continuous variable with a Gaussian distribution, it is possible to formulate a one-dimensional measure of quality — the so-called Bayesian correlation score (BCS) — that is consistent with the sufficiency relation. That is, the values of the BCS order forecasting systems by their relative quality and value (i.e., a larger score indicates higher quality and greater value). In situations in which these conditions and assumptions are satisfied, the BCS obviously offers advantages over alternative measures as a basis for comparative verification. It remains to be seen whether or not it is possible to formulate other one-dimensional measures that, under specific conditions and/or assumptions, are consistent with the sufficiency relation. Until such measures have been defined, this concept will remain of limited use as a screening criterion for verification measures.

Propriety. This screening criterion is based on the principle that a verification measure should not encourage forecasters to make forecasts that differ from their true judgments (differences between forecasts and judgments are indicative of "hedging"). It applies to a particular class of verification measures called "scoring rules" (see Section 2.7). Since forecasters' judgments are inherently probabilistic, this principle is applicable only in contexts in which forecasts are expressed in a probabilistic format. Moreover, the fact that the propriety criterion relates to the correspondence between forecasters' judgments and their forecasts implies that it is especially relevant in contexts involving *subjective* probabilistic forecasts.

Consider a situation involving an underlying variable whose range of values has been divided into m mutually exclusive and collectively exhaustive (m.e.c.e.) categories or events. Let $S = S(f, x)$ denote a generic scoring rule with positive orientation (i.e., larger scores are better), and let p_k denote the forecaster's judgment concerning the likelihood of occurrence of the kth event ($k = 1, \ldots, m$). Further, let $S_k(f)$ denote the score assigned (by S) to the forecast f when the kth event ($x = x_k$) occurs and let $E[S(f, p)]$ denote the forecaster's subjective expected score, where

$$E[S(f, p)] = \sum_k p_k S_k(f). \qquad (2.4)$$

$E[S(f, p)]$ in equation (2.4) is a subjective expected score in the sense that it represents the weighted average of the actual scores [i.e., the $S_k(f)$ for $k = 1, \ldots, m$], where the weights are the forecaster's subjective probabilities that she will receive these scores.

The scoring rule $S(f,x)$ is *strictly proper* if $E[S(p,p)] > E[S(f,p)]$ for all $f \neq p$, *proper* if $E[S(p,p)] \geq E[S(f,p)]$ for all f, and *improper* if $E[S(p,p)] < E[S(f,p)]$ for some $f \neq p$ (Murphy and Daan, 1985; Winkler and Murphy, 1968). Thus, a scoring rule is strictly proper if the maximum expected score can be achieved only when the forecaster makes her forecast f correspond exactly to her judgment p. A scoring rule is proper if forecasts other than $f = p$ lead to the same maximum expected score. Finally, a scoring rule is improper if the maximum expected score is achieved by making a forecast $f \neq p$.

Examples of scoring rules that are strictly proper include the Brier or quadratic score, the logarithmic score, and the spherical score (Winkler and Murphy, 1968). Skill scores based on the Brier score and defined in the usual way (see Section 3.2) are approximately strictly proper for large data samples (Murphy, 1973). The expression for (expected) forecast value in Section 6 is an example of a proper scoring rule. Finally, the so-called linear scoring rule, defined in terms of the absolute difference between forecast probabilities and event occurrence or nonoccurrence (in the case of a dichotomous variable), is an example of an improper scoring rule. This scoring rule leads to an extreme form of hedging, in that forecasters are encouraged to transform their probabilistic judgments into nonprobabilistic forecasts.

Use of the propriety criterion provides a means of screening alternative scoring rules as overall measures of the quality of probabilistic forecasts. Clearly, it is desirable to restrict the choice of such measures to the class of strictly proper scoring rules whenever possible and to avoid the use of improper scoring rules.

Consistency. Although the concept of propriety cannot be applied to nonprobabilistic forecasts, a related but weaker concept called consistency can be invoked in contexts involving such forecasts. This concept is especially applicable in situations involving nonprobabilistic forecasts of continuous variables. It is based on the premise that forecasters follow — or are instructed to follow — a specific directive (or rule) when translating their probabilistic judgments into nonprobabilistic forecasts.

The consistency criterion states that the primary measure used to verify the forecasts should be consistent with the directive followed by the forecaster (Murphy and Daan, 1985). For example, if the directive states "forecast the mean value of your judgmental probability distribution," then the mean square error is an appropriate — that is, consistent — verification measure. This measure is consistent with the directive because the score assigned by the mean square error is minimized by choosing the mean of the distribution as the forecast. On the other hand, if the directive states "forecast the median value of your subjective probability distribution," then the mean absolute error would be a consistent measure. This measure is minimized by forecasting the median of the forecaster's distribution. Thus, the consistency concept can be used to screen alternative measures of forecast quality for nonprobabilistic forecasts, identifying those measures that are — and are not — consistent with the directive given to the forecaster.

Equitability. The concept of equitability applies to verification measures involving nonprobabilistic forecasts of discrete variables. This concept is based on the principle that constant forecasts of any event — as well as forecasts produced by a procedure in which the forecast event is chosen at random — should receive the same expected score (Gandin and Murphy, 1992). According to this principle, the expected scores attained by constant forecasts of events that occur frequently and constant forecasts of events that occur infrequently should be identical.

An example of a measure that satisfies this criterion in dichotomous situations is Kuiper's performance index (see Murphy and Daan, 1985; Wilks, 1995). Verification measures that do not satisfy this criterion include the fraction of correct forecasts, as well the threat score or critical success index. The concept of equitability does not provide a unique solution in the case of polychotomous situations, so that additional assumptions are required to apply this criterion to screen verification measures in such problems.

Application of screening criteria. As previously noted, application of these screening criteria is generally not intended to identify a single best verification measure. The objective is instead to identify those measures that do — and those that do not — satisfy the criteria. Presumably, measures that are judged acceptable on the basis of such screening tests would be preferred to measures that are judged unacceptable. It may be possible to identify other criteria that can be added to the current set, thereby reducing still further the class of acceptable measures. In any case, it is important to keep in mind that no single verification measure can assess all potentially relevant aspects of forecast quality.

3. Absolute verification: methods and applications

Consideration of the problem of absolute verification from a DO perspective leads to the identification of three classes of verification methods: (i) the basic joint, conditional, and marginal distributions themselves; (ii) summary measures of these distributions; and (iii) measures of various aspects of forecast quality. These methods are defined in this section, and their use and interpretation are illustrated by applying the methods to samples of nonprobabilistic maximum temperature (Tmax) forecasts for Minneapolis, Minnesota (Murphy, Brown, and Chen, 1989), and probability of precipitation (PoP) forecasts for St. Louis, Missouri (Murphy and Winkler, 1992).

Both verification data samples include forecasts of two types: so-called objective forecasts produced by numerical-statistical models, and subjective forecasts formulated by U.S. National Weather Service forecasters. Here we focus on the evaluation of various basic aspects of the quality of the objective and subjective forecasts separately. Use of these verification methods to compare the two types of forecasts is described in Section 4.2.

The verification methods and measures defined in this section constitute tailored versions of a body of evaluation methodology that is applicable in principle to all verification data samples. In particular, this set of methods and measures is consistent with the multidimensional structure of verification problems and the multifaceted nature of forecast quality. Brief discussions of the way in which these methods/measures can be tailored to verification problems involving different types of forecasts — as well as overviews of some other verification methods — are included in Section 5.

3.1. Basic distributions and summary measures

In view of the fact that the joint distribution $p(f, x)$ contains all the information relevant to forecast quality (see Section 2.2), absolute verification should begin with an examination of this distribution. Evaluation of $p(f, x)$ provides overall insight into the relationship between forecasts and observations. The joint distribution can be depicted in several different ways. For example, it can be displayed graphically in the form of a scatter diagram or a bivariate histogram or numerically in the form of a contingency table. Bivariate histograms for samples of 24-hour nonprobabilistic Tmax forecasts are presented in Figure 2.1. In this case, strong relationships between f and x appear to exist for both types of forecasts. Further insight can be obtained from careful examination of these diagrams. For example, Figure 2.1a reveals a tendency toward overforecasting in the case of the objective forecasts (i.e., forecast temperatures exceed observed temperatures more often than vice versa).

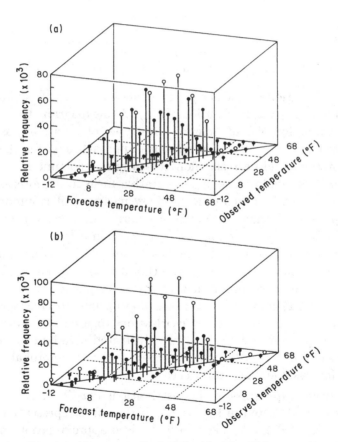

Figure 2.1. Bivariate histograms displaying $p(f, x)$ for 24-hour maximum temperature forecasts in the winter season for Minneapolis, Minnesota (open circles represent case of $f = x$): (a) objective forecasts; (b) subjective forecasts. (From Murphy, Brown, and Chen, 1989)

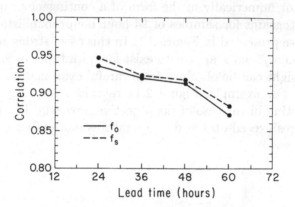

Figure 2.2. Line diagram displaying ρ_{fx} as a function of lead time for objective (f_o) and subjective (f_s) maximum temperature forecasts in the winter season for Minneapolis, Minnesota. (From Murphy, Brown, and Chen, 1989)

The correlation coefficient ρ_{fx} is a (one-dimensional) summary measure of the joint distribution $p(f, x)$. A line diagram displaying ρ_{fx} as a function of lead time for both types of Tmax forecasts is presented in Figure 2.2. Correlations approach 0.95 for the 24-hour forecasts and decrease monotonically as lead time increases. When ρ_{fx} is employed as a quantitative measure of forecasting performance (as opposed to a summary measure), it is important to keep in mind that this measure ignores any unconditional or conditional biases in the forecasts (see Section 3.2).

Information regarding various basic aspects of forecast quality can be obtained by examining the conditional and marginal distributions. In view of the factorizations of $p(f, x)$ set forth in equations (2.1) and (2.2), it seems appropriate to consider the components of the respective factorizations together. The insights provided by these distributions are illustrated here by evaluating samples of PoP forecasts, as well as the samples of nonprobabilistic Tmax forecasts already introduced.

Conditional quantiles of the distributions $q(x|f)$ for the Tmax forecasts are depicted in Figure 2.3. If the curve representing the conditional medians is taken to be a close approximation to the conditional means (a reasonable assumption in this case), comparison of this curve with the 45° line provides insight into the reliability of these forecasts. In this regard, the objective forecasts (Figure 2.3a) exhibit some overforecasting, at least for relatively

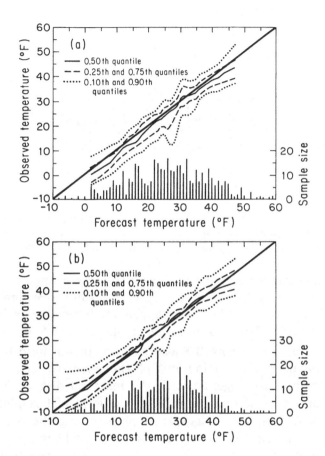

Figure 2.3. Quantiles of conditional distributions $q(x|f)$ (above), and marginal distribution $s(f)$ (below), for 24-hour maximum temperature forecasts in the winter season for Minneapolis, Minnesota: (a) objective forecasts; (b) subjective forecasts. (From Murphy, Brown, and Chen, 1989)

high and low Tmax forecasts. The subjective forecasts (Figure 2.3b) appear to be quite reliable over the entire range of forecast temperatures. Similar diagrams can be produced for the conditional distributions $r(f|x)$; they are omitted here because of space considerations.

The spread of the conditional quantiles (i.e., the difference between the 0.25th and 0.75th quantiles, or between the 0.10th and 0.90th quantiles) provides insight into the accuracy of the Tmax forecasts as a function of the forecast temperature. This spread

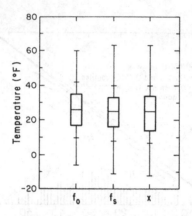

Figure 2.4. Box plots of marginal distributions for 24-hour maximum temperature forecasts — and the corresponding observations — in the winter season for Minneapolis, Minnesota. (From Murphy, Brown, and Chen, 1989)

exhibits no clear pattern of behavior in the case of the objective forecasts (Figure 2.3a). In the case of the subjective forecasts (Figure 2.3b), however, the spread appears to be smaller for intermediate forecasts than it is for relatively high or low forecasts.

The diagrams in Figure 2.3 also contain, in the form of bar charts, the marginal distributions of the forecasts [i.e., $s(f)$]. These distributions, together with the marginal distribution of observations, $t(x)$, are summarized in the form of box plots in Figure 2.4. Among other things, these plots confirm the existence of a bias in the objective Tmax forecasts (initially identified in the bivariate histogram). In addition, the box plots indicate that the variability of the observations exceeds the variability of both types of forecasts.

Conditional distributions $q(x|f)$ and marginal distributions $s(f)$ for matched samples of PoP forecasts are depicted in Figure 2.5. Since these forecasts relate to a binary variable (i.e., $x = 1$ or $x = 0$), each conditional distribution $q(x|f)$ contains only one independent probability [i.e., $q(x = 1|f) + q(x = 0|f) = 1$ for each f]. As a result, it is reasonable to begin the verification process in this case with the conditional and marginal distributions (rather than the joint distribution). Moreover, it should be noted that $q(x = 1|f) = \mu_{x|f}$ in the case of these PoP forecasts.

Examination of the lower part of Figure 2.5 reveals that both types of forecasts are quite reliable; that is, the conditional rela-

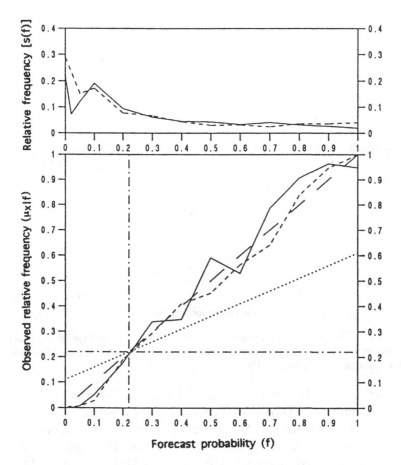

Figure 2.5. Reliability diagram (bottom) and sharpness diagram (top) for 12-to 24-hour objective (solid line) and subjective (dashed line) PoP forecasts in the cool season for St. Louis, Missouri. (From Murphy and Winkler, 1992)

tive frequencies $q(x = 1|f)$ defining the empirical reliability curve correspond quite closely to the forecast probabilities f over the entire range of forecasts. The upper part of Figure 2.5 depicts the marginal distributions $s(f)$, which describe the sharpness (or refinement) of the forecasts. Sharp (refined) forecasts are characterized by u-shaped distributions, with probabilities near zero and one used relatively frequently and intermediate probabilities used relatively infrequently. Although these PoP forecasts use probabilities less than or equal to 0.2 quite frequently, it is evident that neither the objective nor subjective forecasts are very sharp.

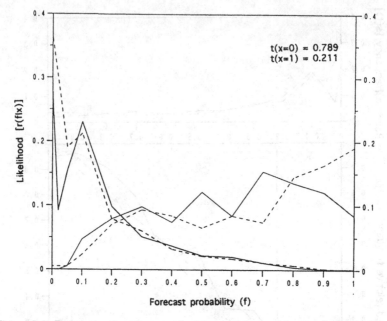

Figure 2.6. Discrimination diagram for 12- to 24-hour objective (solid line) and subjective (dashed line) PoP forecasts in the cool season for St. Louis, Missouri. (From Murphy and Winkler, 1992)

Insight into other aspects of quality can also be obtained from the diagrams in Figure 2.5. The horizontal line $\mu_{x|f} = \mu_x$ represents no resolution (see Section 2.5), and the line intermediate between this horizontal line and the 45° line (which represents perfect reliability) represents no skill (Hsu and Murphy, 1986). This no-skill line is defined in terms of a skill score based on the mean square error (see equation 2.7), in which the standard of reference is a constant forecast of the sample climatological probability (see Section 3.2). Subsamples of forecasts possess positive skill if the corresponding points $(\mu_{x|f}, f)$ lie to the right (left) of the vertical line $f = \mu_x$ and above (below) the no-skill line. Comparison of the empirical reliability curves with the horizontal line $\mu_{x|f} = \mu_x$, keeping in mind the distribution $s(f)$, indicates that both types of forecasts possess substantial resolution. Moreover, most if not all of the points defining these curves lie in regions of positive (subsample) skill, thereby implying that the overall skill of the forecasts should be strongly positive (see Section 3.2).

Table 2.2. Summary measures of marginal distributions $s(f)$ and $t(x)$ and joint distribution $p(f, x)$ for objective (f_o) and subjective (f_s) PoP forecasts in the cool season for St. Louis, Missouri

Lead time (h)	Type of forecast	Means		Variances		Correlation coefficient	Sample size
		μ_f	μ_x	σ_f^2	σ_x^2	ρ_{fx}	n
12–24	f_o	0.220	0.211	0.074	0.166	0.722	1,021
12–24	f_s	0.230	0.211	0.089	0.166	0.755	1,021
24–36	f_o	0.234	0.222	0.067	0.173	0.643	1,019
24–36	f_s	0.227	0.222	0.075	0.173	0.615	1,019
36–48	f_o	0.208	0.212	0.047	0.167	0.565	1,007
36–48	f_s	0.209	0.212	0.054	0.167	0.543	1,007

Source: Adapted from Murphy and Winkler, 1992.

Conditional distributions $r(f|x)$ and marginal distributions $t(x)$ for the PoP forecasts are depicted in Figure 2.6. This figure contains the likelihoods $r(f|x = 0)$ and $r(f|x = 1)$. Although these distributions overlap, substantial discrimination exists for both types of forecasts. That is, high probabilities are used much more frequently than low probabilities when precipitation occurs $(x = 1)$, and low probabilities are used much more frequently than high probabilities when precipitation does not occur $(x = 0)$. Since PoP forecasts relate to a binary event, the marginal distribution of the observations consists simply of $t(x = 1)$ and $t(x = 0)$, where $t(x = 1) + t(x = 0) = 1$.

Summary measures of the joint and marginal distributions for the PoP forecasts and observations are included in Table 2.2. These measures indicate a slight tendency toward overforecasting for both types of forecasts (i.e., $\mu_f > \mu_x$) for the 12- to 24-hour and 24- to 36-hour lead times. Moreover, the variances of the forecasts are considerably smaller than the variances of the observations for all three lead times (in this regard, recall that the forecasts are probabilities, whereas the observations are binary variables). The correlation coefficient, a summary measure of $p(f, x)$, indicates that, as expected, the association between forecasts and observations tends to decrease as lead time increases.

Table 2.3. Summary measures of conditional distributions of forecasts given observations, $r(f|x)$, for objective (f_o) and subjective (f_s) PoP forecasts in the cool season for St. Louis, Missouri

Lead time (h)	Type of forecast	Means		Variances		Sample sizes	
		$\mu_{f|x=0}$	$\mu_{f|x=1}$	$\sigma^2_{f|x=0}$	$\sigma^2_{f|x=1}$	$n_{x=0}$	$n_{x=1}$
12–24	f_o	0.118	0.598	0.026	0.070	806	215
12–24	f_s	0.114	0.665	0.027	0.080	806	215
24–36	f_o	0.145	0.547	0.031	0.068	793	226
24–36	f_s	0.137	0.542	0.036	0.083	793	226
36–48	f_o	0.144	0.445	0.026	0.055	794	213
36–48	f_s	0.144	0.453	0.032	0.061	794	213

Source: Adapted from Murphy and Winkler, 1992.

Table 2.3 contains summary measures of the conditional distributions of forecasts given observations, $r(f|x)$. The conditional means, $\mu_{f|x=0}$ and $\mu_{f|x=1}$, for both types of PoP forecasts generally increase and decrease, respectively, as lead time increases. Thus, the difference between these means, which represents an overall measure of discrimination, decreases as lead increases. The conditional variances indicate that the variability of the forecasts when precipitation occurs $(x = 1)$ is considerably greater than the variability of the forecasts when precipitation does not occur $(x = 0)$. Since the means associated with the conditional distributions of observations given forecasts, $q(x|f)$, together with the forecasts themselves, define the points on the empirical reliability curves (see Figure 2.5), these conditional means are not presented here in tabular form.

3.2. Some basic measures of aspects of quality

In this section we define basic quantitative measures of the various aspects of forecast quality. These performance measures include the mean error, the mean square error, and a skill score based on the mean square error, as well as terms in decompositions of the mean square error and skill score. To illustrate the interpretation of these measures, they are then applied to the verification data samples — containing Tmax and PoP forecasts — considered previously.

The mean error (ME) of the forecasts in the verification data sample is defined here as the difference between the mean forecast μ_f and the mean observation μ_x. That is,

$$\mathrm{ME}(f, x) = \mu_f - \mu_x. \tag{2.5}$$

The ME is a measure of (unconditional or systematic) bias, with positive (negative) values of ME being indicative of overforecasting (underforecasting). When ME = 0, the forecasts are (unconditionally) unbiased. In some contexts, overall bias is measured in terms of the ratio of — rather than the difference between — these means (in this case, $\mu_f/\mu_x = 1$ represents unbiased forecasts).

The mean square error (MSE) of the forecasts in the verification data sample can be defined as follows:

$$\mathrm{MSE}(f, x) = \sum_f \sum_x p(f, x)(f - x)^2. \tag{2.6}$$

The MSE is a measure of accuracy (see Section 2.6), and its values are nonnegative [MSE $\geq$ 0, with equality only when $p(f, x) = 0$ for all $f \neq x$]. Smaller values of the MSE are indicative of greater accuracy. As expressed in equation (2.6), the MSE makes no distinction between nonprobabilistic and probabilistic forecasts. In the meteorological literature, it is common practice to refer to the mean square error of probabilistic forecasts as the Brier score (BS) (Brier, 1950). We will make use of both the MSE and BS notations here.

Skill scores (SSs) are usually defined as the fractional (or percentage) improvement in the accuracy of the forecasts of interest over the accuracy of forecasts based on a naive forecasting method such as climatology or persistence (e.g., Brier and Allen, 1951; Murphy and Daan, 1985). Thus, if the MSE is taken as the measure of accuracy, then the SS$_{\mathrm{MSE}}$ can be defined as follows:

$$\mathrm{SS}_{\mathrm{MSE}}(f, r, x) = 1 - [\mathrm{MSE}(f, x)/\mathrm{MSE}(r, x)], \tag{2.7}$$

where $\mathrm{MSE}(r, x)$ is the MSE for forecasts r based on the naive standard of reference R [note that $\mathrm{MSE}(f, x) = 0$ for perfect forecasts]. Values of the SS$_{\mathrm{MSE}}$ are positive if $\mathrm{MSE}(f, x) < \mathrm{MSE}(r, x)$, zero if $\mathrm{MSE}(f, x) = \mathrm{MSE}(r, x)$, and negative if $\mathrm{MSE}(f, x) > \mathrm{MSE}(r, x)$.

Since $\mathrm{MSE}(f, x) = 0$ when $p(f, x) = 0$ for all $f \neq x$, the value of the $\mathrm{SS}_{\mathrm{MSE}}$ for perfect forecasts is unity.

Climatology and persistence are the standards of reference most often used when skill scores are defined. In the case of climatology, the mean observation (or probability of the event) based on the verification data sample itself is a particularly convenient choice as a standard of reference. In those situations in which the verification data base is relatively large, this sample mean or probability (denoted by μ_x) should be approximately equal to the long-term historical climatological mean or probability. (It should be noted, however, that the use of sample climatology as a standard of reference fails to give the forecasting system or forecaster credit for recognizing differences between sample and long-term climatological probabilities.) The MSE of forecasts based solely on this mean or probability is $\mathrm{MSE}(\mu_x, x) = \sigma_x^2$, in which case equation (2.7) becomes

$$\mathrm{SS}_{\mathrm{MSE}}(f, \mu_x, x) = 1 - [\mathrm{MSE}(f, x)/\sigma_x^2]. \qquad (2.8)$$

Under this assumption, skill is positive (negative) when the MSE of the forecasts is less (greater) than the variance of the observations.

The values of ME (bias) in equation (2.5), MSE (accuracy) in equation (2.6), and $\mathrm{SS}_{\mathrm{MSE}}$ (skill) in equation (2.8) for the PoP forecasts for St. Louis, Missouri, are presented in Table 2.4. Overall, the forecasts exhibit relatively little bias. The accuracy and skill of the forecasts are greater in the cool season than in the warm season. Both forecast accuracy and forecast skill decrease as lead time increases. Comments related to the relative performance of the objective and subjective forecasts are reserved for Section 4.2.

The MSE can be decomposed in several different ways, with the terms in these decompositions representing quantitative measures of various aspects of quality. A basic decomposition of the MSE is

$$\mathrm{MSE}(f, x) = (\mu_f - \mu_x)^2 + \sigma_f^2 + \sigma_x^2 - 2\sigma_f\sigma_x\rho_{fx} \qquad (2.9)$$

(Murphy, 1988). This decomposition is easily motivated by reference to the familiar statistical expression for the variance of a difference; that is, $\mathrm{Var}(f - x) = \sigma_{f-x}^2 = E[(f - x)^2] - [E(f - x)]^2$. In equation (2.9), a measure of accuracy (MSE) has been decomposed into a measure of unconditional bias $[(\mu_f - \mu_x)^2]$, a measure of sharpness (or variability) (σ_f^2), a measure of uncertainty (σ_x^2),

Table 2.4. Overall measures of bias (ME), accuracy (BS), and skill (SS$_{BS}$) for objective (f_o) and subjective (f_s) PoP forecasts for St. Louis, Missouri

Lead time (h)	Type of forecast	Sample size n	Mean error (ME)	Brier score (BS)	Skill score (SS$_{BS}$)
(a) Cool season					
12–24	f_o	1,021	0.009	0.080	0.517
12–24	f_s	1,021	0.020	0.072	0.567
24–36	f_o	1,019	0.013	0.101	0.412
24–36	f_s	1,019	0.005	0.108	0.376
36–48	f_o	1,007	−0.004	0.114	0.318
36–48	f_s	1,007	−0.002	0.118	0.295
(b) Warm season					
12–24	f_o	993	0.020	0.110	0.321
12–24	f_s	993	0.021	0.103	0.365
24–36	f_o	990	−0.014	0.133	0.228
24–36	f_s	990	−0.006	0.130	0.240
36–48	f_o	980	−0.017	0.131	0.196
36–48	f_s	980	−0.013	0.129	0.210

Source: Adapted from Murphy and Winkler, 1992.

and a measure of association ($2\sigma_f\sigma_x\rho_{fx} = 2\sigma_{fx}$, where σ_{fx} denotes the covariance of the forecasts and observations). The last three terms on the right-hand side of equation (2.9), taken together, constitute the variance of the forecast errors (i.e., σ^2_{f-x}). This decomposition is applicable to all verification data samples, regardless of the nature (or treatment) of the underlying variable and the format of the forecasts.

The terms in this basic decomposition for the Tmax forecasts are presented in Table 2.5. Some bias is exhibited by the objective forecasts (1.8–2.4°F), whereas the subjective forecasts are almost completely unbiased. The variability of both types of forecasts is less than the variability of the observations. Moreover, the variability of the objective and subjective forecasts — and the

Table 2.5. Decomposition of MSE(f, x) for objective (f_o) and subjective (f_s) maximum temperature forecasts in the winter season for Minneapolis, Minnesota

Lead time (h)	Type of forecast	Sample size n	MSE	$(\mu_f - \mu_x)^2$	σ_f^2	σ_x^2	$2\sigma_f\sigma_x\rho_{fx}$
24	f_o	417	24.9	3.2	148.6	174.9	302.0
24	f_s	417	18.0	0.0	154.0	174.9	310.8
36	f_o	405	34.4	5.8	149.3	184.1	304.8
36	f_s	405	26.9	0.1	154.4	184.1	311.6
48	f_o	416	33.9	4.4	143.8	177.8	292.0
48	f_s	416	28.4	0.0	137.9	177.8	287.4
60	f_o	397	49.6	5.3	129.8	182.7	268.2
60	f_s	397	40.5	0.0	129.6	182.7	271.8

Source: Adapted from Murphy, Brown, and Chen, 1989.

covariability between the respective forecasts and the observations — generally decreases as lead time increases.

Other decompositions of the MSE can be formulated by conditioning on either the forecast f or the observation x. These decompositions are related to the CR and LBR factorizations of the joint distribution $p(f, x)$ (see Section 2.3). Conditioning on the forecast f leads to the following CR decomposition of the MSE:

$$\text{MSE}_{\text{CR}}(f, x) = \sigma_x^2 + E_f(\mu_{x|f} - f)^2 - E_f(\mu_{x|f} - \mu_x)^2, \quad (2.10)$$

where E_f denotes an expectation with respect to the distribution of forecasts [i.e., a weighted average — using weights $s(f)$ — over all forecasts]. The expression in equation (2.10) represents a decomposition of a measure of accuracy (MSE) into a measure of uncertainty (σ_x^2), a measure of reliability (or type 1 conditional bias) $[E_f(\mu_{x|f} - f)^2]$, and a measure of resolution $[E_f(\mu_{x|f} - \mu_x)^2]$.

The measure of reliability in equation (2.10) is simply the weighted squared deviation of the points defining the empirical reliability curve from the 45° line representing perfect reliability (see Figure 2.5), where the weights are the probabilities that constitute the marginal distribution of forecasts, $s(f)$. This nonnegative term

Table 2.6. Decomposition of $MSE_{CR}(f,x)$ related to the calibration-refinement factorization of $p(f,x)$ for objective (f_o) and subjective (f_s) PoP forecasts in the cool season for St. Louis, Missouri

Lead time (h)	Type of forecast	Sample size n	MSE_{CR}	σ_x^2	$E_f(\mu_{x|f}-f)^2$	$E_f(\mu_{x|f}-\mu_x)^2$
12–24	f_o	1,021	0.080	0.166	0.002	0.088
12–24	f_s	1,021	0.072	0.166	0.002	0.096
24–36	f_o	1,019	0.101	0.173	0.002	0.073
24–36	f_s	1,019	0.108	0.173	0.002	0.067
36–48	f_o	1,007	0.114	0.167	0.002	0.055
36–48	f_s	1,007	0.118	0.167	0.003	0.052

Source: Adapted from Murphy and Winkler, 1992.

vanishes only for completely reliable forecasts (i.e., $\mu_{x|f} = f$ for all f). The measure of resolution in equation (2.10) is simply the weighted squared deviation of the points defining the same empirical (reliability) curve from the horizontal line representing no resolution (i.e., $\mu_{x|f} = \mu_x$ for all f), where the weights are once again the components of the marginal distribution $s(f)$. Since it is preceded by a negative sign, larger values of this nonnegative term are indicative of greater resolution.

The results of applying this decomposition of the MSE (or BS) to the PoP forecasts are summarized in Table 2.6. It is evident that the lack of complete reliability, noted in Section 3.1, contributes very little to the MSE for either the objective or subjective forecasts. The resolution term is substantially larger than the reliability term for both types of forecasts, and it decreases as lead time increases.

Conditioning on the observation x leads to the following LBR decomposition of the MSE:

$$MSE_{LBR}(f,x) = \sigma_f^2 + E_x(\mu_{f|x}-x)^2 - E_x(\mu_{f|x}-\mu_f)^2, \quad (2.11)$$

where E_x denotes an expectation with respect to the distribution of observations. The expression in equation (2.11) represents a decomposition of a measure of accuracy (MSE) into a measure of

Table 2.7. Decomposition of $\mathrm{MSE}_{\mathrm{LBR}}(f,x)$ related to the likelihood-base rate factorization of $p(f,x)$ for objective (f_o) and subjective (f_s) PoP forecasts in the cool season for St. Louis, Missouri

Lead time (h)	Type of forecast	Sample size n	$\mathrm{MSE}_{\mathrm{LBR}}$	σ_f^2	$E_x(\mu_{f|x} - x)^2$	$E_x(\mu_{f|x} - \mu_f)^2$
12–24	f_o	1,021	0.080	0.074	0.045	0.038
12–24	f_s	1,021	0.072	0.089	0.034	0.050
24–36	f_o	1,019	0.101	0.067	0.062	0.028
24–36	f_s	1,019	0.108	0.075	0.061	0.028
36–48	f_o	1,007	0.114	0.047	0.081	0.015
36–48	f_s	1,007	0.118	0.054	0.080	0.016

Source: Adapted from Murphy and Winkler, 1992.

sharpness (σ_f^2), a measure of type 2 conditional bias $[E_x(\mu_{f|x}-x)^2]$, and a measure of discrimination $[E_x(\mu_{f|x} - \mu_f)^2]$. These latter two terms — and their signs — indicate that it is desirable for the conditional mean forecasts, $\mu_{f|x=1}$ and $\mu_{f|x=0}$, to approach the respective observations $x = 1$ and $x = 0$ as closely as possible (to decrease type 2 conditional bias) and, at the same time, for these conditional means to differ as much as possible from the overall unconditional mean forecast μ_f (to increase discrimination).

Application of the LBR decomposition of the MSE (or BS) to the PoP forecasts yields the results summarized in Table 2.7. The term that measures type 2 conditional bias (the mean squared difference between $\mu_{f|x}$ and x averaged over all x) and contributes positively to the magnitude of the MSE increases as lead time increases. On the other hand, the term that measures discrimination (the mean squared difference between $\mu_{f|x}$ and μ_f averaged over all x) and contributes negatively to the magnitude of the MSE decreases as lead time increases. Comparisons of the terms in this decomposition for the two types of forecasts are considered in Section 4.2.

A decomposition of the MSE-based skill score in equation (2.7) can be obtained by substituting the basic decomposition of the MSE in equation (2.9) into the expression for $\mathrm{SS}_{\mathrm{MSE}}$ in equation (2.8). After rearranging terms and completing a square, it can be

Table 2.8. Decomposition of $SS_{MSE}(f, \mu_x, x)$ for objective (f_o) and subjective (f_s) maximum temperature forecasts in the winter season for Minneapolis, Minnesota

Lead time (h)	Type of forecast	Sample size n	SS_{MSE}	ρ_{fx}^2	$[\rho_{fx} - (\sigma_f/\sigma_x)]^2$	$[(\mu_f - \mu_x)/\sigma_x]^2$
24	f_o	417	0.858	0.876	0.000	0.018
24	f_s	417	0.897	0.897	0.000	0.000
36	f_o	405	0.813	0.846	0.001	0.031
36	f_s	405	0.854	0.854	0.000	0.000
48	f_o	416	0.809	0.834	0.000	0.025
48	f_s	416	0.840	0.843	0.001	0.000
60	f_o	397	0.728	0.759	0.001	0.029
60	f_s	397	0.778	0.780	0.002	0.000

Source: Adapted from Murphy, Brown, and Chen, 1989.

seen that

$$SS_{MSE}(f, \mu_x, x) = \rho_{fx}^2 - [\rho_{fx} - (\sigma_f/\sigma_x)]^2 - [(\mu_f - \mu_x)/\sigma_x]^2. \quad (2.12)$$

The terms on the right-hand side of equation (2.12) are all nonnegative, and they can be interpreted by reference to a linear regression model in which the forecasts are regressed on the observations (e.g., see Murphy and Winkler, 1992). In this context, the first term is the square of the correlation coefficient ρ_{fx}, a measure of (linear) association between f and x. As noted in Section 2.6, ρ_{fx}^2 represents the fraction of the variability in the observations accounted for (or "explained") by the forecasts.

Forecasts are completely reliable in the context of the regression model only when the regression line possesses zero intercept *and* unit slope (i.e., only when it coincides with the 45° line). Under this condition, $\sigma_f = \rho_{fx}\sigma_x$, and it follows that the second term on the right-hand side of equation (2.12) is a measure of reliability (calibration, conditional bias in the type 1 sense). This term vanishes for completely reliable forecasts, and otherwise acts to reduce the skill score. Since the third term is the squared difference between μ_f and μ_x, scaled by the square of σ_x, it is a measure of

Table 2.9 Decomposition of $\text{SS}_{\text{MSE}}(f, \mu_x, x)$ for objective (f_o) and subjective (f_s) PoP forecasts in the cool season for St. Louis, Missouri

Lead time (h)	Type of forecast	Sample size n	SS_{MSE}	ρ_{fx}^2	$[\rho_{fx} - (\sigma_f/\sigma_x)]^2$	$[(\mu_f - \mu_x)/\sigma_x]^2$
12–24	f_o	1,021	0.517	0.521	0.003	0.000
12–24	f_s	1,021	0.567	0.570	0.001	0.002
24–36	f_o	1,019	0.412	0.414	0.000	0.001
24–36	f_s	1,019	0.376	0.378	0.002	0.000
36–48	f_o	1,007	0.318	0.319	0.001	0.000
36–48	f_s	1,007	0.295	0.295	0.001	0.000

Source: Adapted from Murphy and Winkler, 1992.

overall (or unconditional) bias. It vanishes for completely unbiased forecasts, and otherwise acts to reduce the skill score.

Examination of the decomposition in equation (2.12) reveals that SS_{MSE} and ρ_{fx}^2 are equal when the forecasts are conditionally unbiased. As noted in Section 2.6, a verification data sample that is conditionally unbiased for all forecasts is also unconditionally unbiased, but not necessarily vice versa. In this sense, ρ_{fx}^2 can be viewed as a measure of potential skill (Murphy and Epstein, 1989).

The results of applying this decomposition of the SS to the Tmax and PoP forecasts are summarized in Tables 2.8 (p. 53) and 2.9, respectively. Since these samples of forecasts are relatively reliable, the terms measuring type 1 conditional bias (or reliability) and unconditional (or systematic) bias are quite small in most cases. The only exception to this general result is the sample of objective Tmax forecasts, which exhibit a small but not insignificant contribution to the SS from the term that measures unconditional bias. The fact that these two terms are quite small in most cases implies that the values of the SS and ρ_{fx}^2 are similar and that actual skill and potential skill differ very little in most cases for these forecasts.

4. Comparative verification: methods and applications

In this section we address the problem of comparative verification of two or more forecasting systems. Since the process of comparing such systems is transitive, it suffices to compare forecasting systems on a pairwise basis. Section 4.1 discusses the use of methods based on the sufficiency relation as a means of screening alternative forecasting systems. The application of the basic set of verification methods and measures introduced in Section 3.2 to the problem of comparative verification is illustrated in Section 4.2.

4.1. Screening forecasting systems: the sufficiency relation

The conditions under which one forecasting system can be unambiguously judged to be better, in terms of both quality and value, than another forecasting system are embodied in the *sufficiency relation*. In the context of matched comparative verification, forecasting system F is sufficient for forecasting system G if and only if a stochastic transformation $h(g|f)$ exists such that

$$\sum_f h(g|f) r_F(f|x) = r_G(g|x) \text{ for all } x \qquad (2.13)$$

(e.g., DeGroot and Fienberg, 1982; Ehrendorfer and Murphy, 1988). The function $h(g|f)$ qualifies as a stochastic transformation if $0 \le h(g|f) \le 1$ for all f and g and $\sum_g h(g|f) = 1$ for each f. Note that, under the assumption that the marginal distribution $t(x)$ is known, the likelihoods $r_F(f|x)$ and $r_G(g|x)$ in equation (2.13) provide a complete description of the quality of F's and G's forecasts, respectively (see equation 2.2).

Since the sufficiency relation defined by equation (2.13) implies that G's likelihoods can be obtained by an auxiliary randomization of F's likelihoods, the former obviously contain greater uncertainty than the latter. The importance of the sufficiency relation resides in the fact that if it can be shown that F's forecasts are sufficient for G's forecasts, then it follows that F's forecasts are of greater value than G's forecasts to all users regardless of the structure of their individual payoff (or loss) functions.

The sufficiency relation was developed originally by Blackwell (1953) in the context of statistical experiments. It was then applied to information systems by Marschak (1971) and subsequently introduced into the forecasting literature by DeGroot and Fienberg (1982). The sufficiency relation is a quasi-order (Krzysztofowicz and Long, 1991a), in the sense that the comparison of systems F and G can lead to any of three possible results: (i) F is sufficient for G; (ii) G is sufficient for F; or (iii) F and G are insufficient for each other. Result (iii) implies that no function $h(g|f)$ or $h(f|g)$ can be found that satisfies the conditions for a stochastic transformation. In this case, some users may prefer F's forecasts and other users may prefer G's forecasts. The conditions under which forecasting systems are sufficient and insufficient for each other clearly is an important issue in applications of the sufficiency relation to problems of comparative verification.

As a method of screening alternative forecasting systems, the sufficiency relation can be used to determine whether or not a particular forecasting system is dominated by another forecasting system (Clemen, Murphy, and Winkler, 1995). However, as a quasi-order, the method will not always be able to identify the "better" system. Nevertheless, in the context of comparative verification, this relation leads to coherent methods of comparing alternative forecasting systems, eliminating those systems whose forecasts are dominated by the forecasts of other systems, and thereby reducing the number of alternative systems that must be considered. Various applications of the sufficiency relation as a screening method are described briefly in this section.

One screening method based on the sufficiency relation involves the direct search for auxiliary randomizations that satisfy the conditions for a stochastic transformation. This approach was taken by Ehrendorfer and Murphy (1988, 1992a) in assessing the relative quality and value of primitive weather and climate forecasts. For example, Ehrendorfer and Murphy (1988) investigated the conditions under which one forecasting system is sufficient for another forecasting system in the case of nonprobabilistic forecasts for a dichotomous variable (i.e., $f = 0$ or 1 and $x = 0$ or 1). These conditions are described graphically in the two-dimensional sufficiency diagram depicted in Figure 2.7, in which the conditional probabilities $q_{11} = \Pr(x = 1|f = 1)$ and $q_{10} = \Pr(x = 1|f = 0)$ represent the coordinate axes (equivalent descriptions can be formulated in terms of likelihoods or joint probabilities; see Ehrendorfer and Murphy, 1988).

In Figure 2.7 system F is taken as the reference system, with $q_{11}(F) = 0.571$ and $q_{10}(F) = 0.276$ [the sample climatological probability $t_1 = \Pr(x = 1) = 0.4$, the same for all systems under consideration here]. These conditional probabilities completely determine the geometry of the sufficiency diagram, which is defined by horizontal and vertical lines passing through the point (q_{11}, q_{10}) $= (0.571, 0.276)$ in this framework. Three regions are identified: (i) regions S, which consist of all systems G (e.g., $G1$) for which system F is sufficient; (ii) regions S', which consist of all systems G (e.g., $G3$) that are sufficient for system F; and (iii) regions I, which consist of all systems G (e.g., $G2$) that are insufficient for system F.

The constraints that the conditional probabilities of the respective forecasting systems must satisfy to fall in these regions are summarized in Table 2.10. In essence, for an alternative system G to be sufficient for the reference system F, the interval defined by the former's conditional probabilities $q_{11} = \Pr(x = 1|f = 1)$ and $q_{10} = \Pr(x = 1|f = 0)$ must "straddle" the interval defined by the latter's conditional probabilities. Conversely, if F's interval straddles G's interval, then F's forecasts are sufficient for G's forecasts. If the intervals overlap, then systems F and G are insufficient for each other. This latter condition implies that (i) the quality of F's forecasts is, at the same time, superior and inferior to the quality of G's forecasts in one or more respects and (ii) the user population includes at least some individuals who prefer each system's forecasts. In this case neither F nor G can be said to dominate the other system.

The dashed lines in Figure 2.7 are isopleths of the expected Brier score (EBS) [see Ehrendorfer and Murphy (1988) for a definition of the EBS]. It should be noted that these isopleths generally traverse regions of sufficiency (S or S') and regions of insufficiency (I). Thus, from the relative values of EBS alone it is not possible to determine whether F is sufficient for G (or vice versa) or F and G are insufficient for each other. On the other hand, if $\mathrm{EBS}(F) <$ ($>$) $\mathrm{EBS}(G)$, system G (F) cannot be sufficient for system F (G).

Another screening method based on the sufficiency relation involves the formulation of the *forecast sufficiency characteristic* (FSC) (Krzysztofowicz and Long, 1991a). This method, which is based on a theorem presented by Blackwell and Girshick (1954), is applicable only in situations involving dichotomous variables (i.e.,

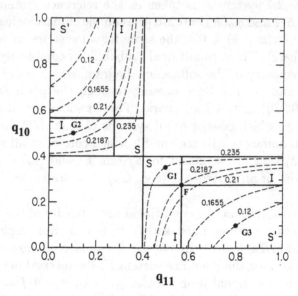

Figure 2.7. Sufficiency diagram for nonprobabilistic forecasts in a dichotomous situation, described in terms of the conditional probabilities $q_{11} = \Pr(x = 1|f = 1)$ and $q_{10} = \Pr(x = 1|f = 0)$. Isopleths represent lines of equal expected half-Brier score. See text for additional details. (From Ehrendorfer and Murphy, 1988)

$x = 0$ or 1). It involves the construction and comparison of FSCs, which are derived from the likelihoods of the respective forecasting systems [i.e., from $r(f|x = 1)$ and $r(f|x = 0)$].

As an example of an application of this screening method, the FSCs for objective and subjective precipitation probability forecasts for Portland, Oregon, are depicted in Figure 2.8 [the FSCs are denoted by $C(t)$ in this figure]. If the FSC of one forecasting system is *superior* to (i.e., lies everywhere to the left and above) the FSC of the other forecasting system, then the former's forecasts are sufficient for the latter's forecasts. In the case of the cool season forecasts (Figure 2.8a), the FSCs clearly do not satisfy the conditions for sufficiency. However, in the case of the warm season forecasts (Figure 2.8b), the FSC for the subjective forecasts is superior to the FSC for the objective forecasts, implying that the former are sufficient for the latter.

Although experience with the use of FSCs to compare forecasting systems is quite limited, this experience suggests that the FSCs

Table 2.10. Constraints on conditional probabilities of reference forecasting system F, $q_{11}(F) = 0.571$ and $q_{10}(F) = 0.276$, and alternative forecasting systems G, $q_{11}(G)$ and $q_{10}(G)$, associated with the regions S, S', and I in Figure 2.7 [with $t_1 = \Pr(x = 1) = 0.4$]

Region	Constraints	Example
S	$q_{10}(F) \leq q_{10}(G) < t_1 < q_{11}(G) \leq q_{11}(F)$	$G1: q_{11} = 0.50, q_{10} = 0.35$
S'	$q_{10}(G) \leq q_{10}(F) < t_1 < q_{11}(F) \leq q_{11}(G)$	$G3: q_{11} = 0.80, q_{10} = 0.10$
I	$q_{10}(F) \leq q_{10}(G) < t_1 < q_{11}(F) \leq q_{11}(G)$ or $q_{10}(G) \leq q_{10}(F) < t_1 < q_{11}(G) \leq q_{11}(F)$	$G2: q_{11} = 0.10, q_{10} = 0.50$

Note: These constraints define the regions contained in the lower right-hand portion of the unit square in Figure 2.7. The regions in the upper left-hand portion of the unit square represent the situation in which the values of q_{11} and q_{10} are interchanged.

of highly competitive forecasting systems seldom will satisfy the conditions for sufficiency. For example, in 24 different situations (i.e., combinations of location, season, and lead time) considered by Murphy and Ye (1990), sufficiency could be demonstrated in only two situations. Similar results are reported by Krzysztofowicz and Long (1991a). In a companion paper, Krzysztofowicz and Long (1991b) used beta distributions to model the likelihoods associated with the objective and subjective probability forecasts and then compared the respective FSCs derived from the models. This approach served to suppress the sampling variability present in the empirical data sets and, as a result, yielded less ambiguous results. That is, the conditions for sufficiency were met in a substantially larger fraction of the situations. For a discussion of other models of distributions of forecasts and/or observations, see Section 2.5.

The utility of the sufficiency relation as a means of screening alternative forecasting systems is limited because it imposes strict conditions on the relationship between the likelihood functions [i.e., the conditional distributions $r(f|x)$] associated with the respective forecasting systems. The strictness of these conditions arises from the fact that no assumptions are made concerning the

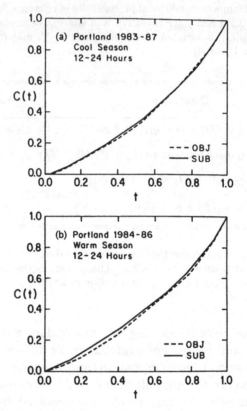

Figure 2.8. FSCs for 12- to 24-hour objective (dashed curve) and subjective (solid curve) PoP forecasts for Portland, Oregon: (a) cool season; (b) warm season. (From Murphy and Ye, 1990)

structure of the payoff or loss functions of the users of the forecasts. Moreover, straightforward application of this relation is difficult in comparative verification problems of moderate or high dimensionality. It may be possible to develop less restrictive — and more easily applicable — versions of the sufficiency relation in situations in which some knowledge is available concerning users' payoff functions. These "tailored" sufficiency relations presumably would be more successful in identifying dominant or dominated forecasting systems in particular situations.

4.2. Distributions-oriented methods and their application

Application of the screening methods described in Section 4.1 may reveal that a forecasting system of potential interest is dominated by another system. In general, however, this screening process will not lead to the identification of a single "best" forecasting system. In such cases, it is necessary to compare the various aspects of forecast quality using methods similar to those described in Section 3.2. To illustrate this approach to comparative verification we will make use of the objective and subjective weather forecasts — and associated results — presented in Section 3. The Tmax forecasts are considered first, and then the PoP forecasts. To conserve space, the objective and subjective forecasts are referred to simply as f_o and f_s, respectively.

Tmax forecasts. From the box plots for the Tmax forecasts and the corresponding observations in Figure 2.4, it is evident that f_o possesses a modest positive (i.e., overforecasting) bias, whereas f_s is relatively unbiased. The magnitude of the respective unconditional biases can be inferred from Table 2.5 — the bias of f_o ranges from 1.8°F to 2.4°F, whereas the bias of f_s is very close to zero. Despite the bias for f_o, the measures of bias in the decomposition of SS_{MSE} detract very little from the overall skill of both types of forecasts (see Table 2.8).

The bivariate histograms in Figure 2.1 provide qualitative insight into the relative association of f_o and f_s for the Tmax forecasts. From a quantitative point of view, the line diagram in Figure 2.2 indicates that ρ_{fx} is slightly higher for f_s than for f_o at all lead times. This difference is consistent with the numerical values of $2\sigma_f\sigma_x\rho_{fx}$ and ρ_{fx}^2 in Tables 2.5 and 2.8, respectively.

The accuracy of the Tmax forecasts, as measured by the MSE, is reported in Table 2.5. According to this measure, f_s is noticeably more accurate than f_o at all lead times. The skill scores for these forecasts in Table 2.8 reveal that f_s is more skillful than f_o across all lead times. In percentage terms, this difference in skill ranges from 3.1% to 5.0%.

Under the assumption that the conditional median temperatures correspond closely to the conditional mean temperatures, the conditional quantile diagrams in Figure 2.3 characterize the reliability of the 24-hour Tmax forecasts. Comparison of these diagrams indicates that the subjective forecasts (Figure 2.3b) are more reliable than the objective forecasts (Figure 2.3a). Despite this apparent difference in reliability, the magnitude of the term in the decomposition of SS_{MSE} that measures reliability is the same for both types of forecasts and detracts very little from the skill score at all lead times (see Table 2.8).

The relative variability of the objective and subjective Tmax forecasts is indicated in Table 2.5. These results reveal that f_s is somewhat more variable than f_o for the 24-hour and 36-hour forecasts, whereas the opposite is true for the 48-hour forecasts.

Information regarding discrimination in the case of the Tmax forecasts has been omitted to conserve space. However, results presented by Murphy et al. (1989) indicate that f_s exhibits greater discrimination than f_o at the 24-hour lead time, but that very little difference in discrimination between the two types of forecasts exists at longer lead times.

PoP forecasts. Comparison of the values of the ME for the objective and subjective PoP forecasts in Table 2.4 (see also Table 2.2) reveals that neither f_o nor f_s possesses any substantial bias. The impact of the unconditional biases that do exist on the skill score SS_{MSE} is nil (see Table 2.9). With regard to the association between the PoP forecasts and observations, the values of ρ_{fx} in Table 2.2 indicate that association is somewhat greater for f_s than for f_o in the case of the 12- to 24-hour forecasts and vice versa in the cases of the 24- to 36-hour and 36- to 48-hour forecasts. Values of ρ_{fx}^2 for both types of forecasts are included in Table 2.9.

The values of the MSE (or BS) for the PoP forecasts in Table 2.4 (see also Tables 2.6 and 2.7) indicate that f_s is slightly more accurate than f_o for the 24-hour forecasts in both seasons. However, at longer lead times, either little if any difference in accuracy exists or f_o slightly exceeds f_s in accuracy (e.g., the 24- to 36-hour forecasts in the cool season). With regard to skill, the skill scores in Table 2.4 (see also Table 2.9) indicate that f_s is noticeably more skillful than f_o for the 12- to 24-hour PoP forecasts in both seasons. Differences in skill are much smaller at longer lead times, with f_o (f_s) possessing slightly higher skill in the cool (warm) season.

Comparison of the empirical reliability curves for the objective and subjective PoP forecasts in Figure 2.5 suggests that the subjective forecasts are more reliable than the objective forecasts. However, the quantitative measure of reliability in the CR decomposition of the MSE (see Table 2.6) reveals little or no difference in overall reliability between f_o and f_s at all lead times. In the case of resolution, the quantitative measure in Table 2.6 indicates that f_s (f_o) is slightly more resolved than f_o (f_s) for the 12- to 24-hour (24- to 36-hour) forecasts. Comparison of the values of σ_f^2 in Table 2.2 (see also Table 2.7) reveals that f_s is sharper than f_o at all lead times, with the difference in sharpness being largest for the 12- to 24-hour forecasts.

The discrimination diagrams for the PoP forecasts in Figure 2.6 suggest that f_s possesses greater discrimination than f_o for the 12- to 24-hour forecasts. This inference is supported by a comparison of the difference in the means of the conditional distributions, $\mu_{f|x=1} - \mu_{f|x=0}$, for the two types of forecasts (see Table 2.3). The respective differences are 0.551 for f_s and 0.480 for f_o. It is interesting to note that the difference in magnitude between these two differences in means is almost entirely due to the fact that $\mu_{f|x=1}$ is much larger for f_s than for f_o (in particular, the difference between the respective values of $\mu_{f|x=0}$ is quite small). At longer lead times the respective differences in these conditional means are similar in magnitude, indicating that the levels of discrimination for the two types of forecasts are about the same. These results are supported by the values of the quantitative measure of discrimination in Table 2.7.

Quality — overall assessment of f_o vis-à-vis f_s. With regard to an overall assessment of the relative quality of the objective and subjective forecasts, it appears that 12- to 24-hour subjective forecasts are of higher quality — in terms of most if not all of its various aspects — than the 12- to 24-hour objective forecasts (this statement applies to both the Tmax and PoP forecasts). However, at longer lead times differences in the various aspects of quality for the two types of forecasts are generally much smaller, and in some cases appear to favor the objective forecasts over the subjective forecasts. In this regard, it should be noted that the results of comparative verification are usually not definitive, in the sense that one forecasting system can be judged unambiguously to be better than another forecasting system. The fact that one system is better than another system on "almost all" aspects of quality does not guarantee that such a strong conclusion is warranted. Only comparisons based on the sufficiency relation can yield unambiguous results regarding the relative quality (and value) of alternative forecasting systems.

5. Other methods and measures

The DO approach and its three classes of methods can be applied to most if not all verification problems. In particular applications, however, it may be desirable (or necessary) to tailor the body of DO methodology to take into account such factors as the nature of the underlying variable and the type of forecasts. For example, the DO methods introduced in Sections 3 and 4 were tailored to nonprobabilistic forecasts of a continuous weather variable in the case of Tmax forecasts and to probabilistic forecasts of a dichotomous weather variable in the case of PoP forecasts. In this section we briefly describe ways in which DO methods can be tailored to verification problems involving nonprobabilistic forecasts of discrete weather variables and probabilistic forecasts of discrete but polychotomous weather variables.

In the case of nonprobabilistic forecasts of discrete variables (e.g., cloud amount, weather types), the basic joint, conditional, and marginal distributions of forecasts and/or observations can be conveniently summarized in the form of contingency tables. Summary measures of the basic distributions — as well as measures of various aspects of forecast quality — can be calculated directly from the joint, conditional, and marginal probabilities that constitute the elements of these tables. In choosing among alternative measures of some aspects of quality, it may be necessary to consider the nature of the underlying weather variable. If the variable is ordinal (i.e., consists of categories with a natural order, such as categories of cloud amount), verification measures that take into account the "distance" between the forecast and observed categories may be appropriate. On the other hand, when the variable is nominal (i.e., consists of events without a natural order, such as weather types), the concept of distance is not meaningful and verification measures need not take this factor into account. A recent discussion of verification methods for nonprobabilistic forecasts of discrete variables can be found in Wilks (1995, pp. 238–250).

Verification problems involving probabilistic forecasts of polychotomous variables are problems of relatively high dimensionality. In the case of nominal variables, such problems can be simplified — with little or no loss of information — by treating the forecasts and observations associated with each event separately. If this approach is followed, then the original polychotomous problem would be replaced by n^x dichotomous problems (i.e., a dichotomous problem associated with each event defining the underlying polychotomous variable), and the body of methods used to evaluate PoP forecasts in Sections 3 and 4 could be applied to the verification data sample associated with each dichotomous problem. A similar approach could be followed in the case of ordinal variables (in this case, the polychotomous problem would be replaced by $n^x - 1$ dichotomous problems, one associated with each threshold defining the categories of the underlying polychotomous variable), but it would generally lead to the loss of information concerning some aspects of forecast quality. Moreover, in the case of ordinal polychotomous variables, the appropriate squared-error measure of overall accuracy is the ranked probability score (RPS) (Epstein, 1969; Murphy, 1971). In addition to taking distance into account, the RPS — and skill scores based on the RPS — can be decomposed into measures of other aspects of quality in a manner analogous to the BS and the SS_{BS} in Section 3.2. The full range of DO methods have yet to be applied to probabilistic forecasts in (unmodified) polychotomous verification problems. For further discussion of verification problems involving probabilistic forecasts, see Murphy and Daan (1985) and Wilks (1995).

Verification methods and measures almost always involve various explicit and/or implicit assumptions, and the body of DO methodology described in this chapter is no exception. For example, the basic measures of accuracy and skill introduced in Section 3.2 are defined in terms of the square of the overall degree of correspondence between forecasts and observations. In effect, it has been assumed here that the importance of differences between forecasts and observations is proportional to the square of the magnitude of these differences. The mean square error criterion seems reasonable in many situations, and it facilitates the processs of decomposing the basic measures into measures of other aspects of forecast quality. However, verification methods and measures of overall accuracy or skill based on other criteria have been proposed. For example, Mielke (1991) advocates the use of a linear measure of correspondence between forecasts and observations; a measure in effect analogous to the mean absolute error. Another approach has been proposed by Potts et al. (1996); it involves measuring the correspondence between forecasts and observations in probability space rather than in the space of values of the underlying variable. In this approach, forecast errors are defined in terms of (linear) differences between the respective cumulative probabilities of the forecasts and observations according to the climatological distribution of the variable, thereby assigning larger penalties to errors in regions of probability space in which forecasts/observations are more likely to occur (see Wilks, 1995, pp. 257–258).

In recent years, verificaton methods based on concepts derived from signal detection theory (SDT) (e.g., Swets, 1988) have been used to assess and compare forecasting performance. The SDT approach to verification problems involving weather forecasts was initially exploited by Mason (1982), who described SDT-based methods in some detail and then reported the results of applying these methods to samples of probabilistic forecasts. The basic tool in the SDT approach is the receiver operating characteristic (ROC) — a curve representing the relationship between $r(f|x = 1)$ (the "probability of detection") versus $r(f|x = 0)$ (the "false alarm ratio") for all possible values of f. As a result, SDT methods are similar in some respects to verification methods associated with the likelihood-base rate factorization of the joint distribution $p(f, x)$ (see Section 2.3). The ROC can be calculated using empirical estimates of the conditional probabilities $r(f|x = 1)$ and $r(f|x = 0)$, or it can be determined using Gaussian models of these conditional probabilities (in this latter case, the SDT approach provides an example of a model-based approach to forecast verification; see Section 2.5). Various measures of overall performance can be derived from the estimated or modeled ROCs. Although SDT methods are applicable to all types of forecasts (including forecasts expressed in qualitative terms), these methods are limited to situations involving dichotomous variables. For a recent discussion and application of SDT methods to forecast verification, see Harvey et al. (1992).

As described in this chapter (see Section 2.1), forecast verification is the process and practice of evaluating weather forecasts at specific points (e.g., geographical locations, grid points). In effect, this process/practice ignores the relationship that may exist between forecasts (and observations) at different points. Thus, the problem of *model verification*, which involves the evaluation of forecasts produced by numerical models in the form of spatial arrays (or two-dimensional fields), presents a somewhat different challenge. This problem generally involves nonprobabilistic forecasts, and it can be approached in at least two quite different ways. Traditionally, model verification has been performed by comparing the forecasts and the corresponding observations (or analyzed values) on a point-by-point basis. If this approach is followed, then the DO methods described in Sections 3 and 4 can be used to assess or compare various aspects of forecast quality. In practice, model verification has usually consisted of calculating one or two overall measures of performance such as the mean square error or the (anomaly) correlation coefficient (e.g., see Wilks, 1995, pp. 272–281).

In the point-by-point approach, the correspondence between coherent features — for example, the phase or amplitude of waves — in the respective spatial arrays is not explicitly considered. Model verification in terms of coherent features — or, more generally, spatial patterns encountered in the two-dimensional fields — is not well-developed and is seldom attempted. Some insight into the extent to which differences between forecasts and observations (or analyzed values), calculated on a point-by-point basis, are due to differences in the phase or amplitude of features can be obtained by appealing to the decomposition of SS_{MSE} in equation (2.12). In this expression, the term ρ_{fx}^2 can be interpreted as the degree of association in phase and the term $[\rho_{fx} - (s_f/s_x)]^2$ can be interpreted as the degree of correspondence in amplitude (in applications of this decomposition to model verification problems, the forecasts and observations are usually expressed as anomalies). Further discussion of this decomposition and its application to model verification problems can be found in Livezey (1995) and Murphy and Epstein (1989). Recent attempts to develop methods of model verification based on spatial patterns or features are reported by Briggs and Levine (1997) and Hoffman et al. (1995).

6. Forecast quality and forecast value

This chapter has focused on methods of assessing forecast quality and its various aspects. However, it is not possible — or necessarily even desirable — to divorce considerations of forecast quality entirely from considerations of forecast value. Even the relatively simple choice of a measure of accuracy — for example, the choice between the mean absolute error and the mean square error — involves considerations related to forecast use and value (for an expository discussion of the principal types of goodness in forecasting and their interrelationships, see Murphy, 1993). Moreover, the primary focus of this book — in particular, Chapters 3 through 6 — is the value of weather and climate forecasts. For these reasons, some basic features of the relationship between forecast quality and forecast value are discussed briefly here. This discussion may help to connect the treatment of forecast quality in this chapter with the treatments of forecast value in subsequent chapters.

It is useful here to distinguish between two types of situations in which forecast use and value are of interest. In the first type of situation all potential users of the forecasts are of interest and nothing is known or assumed about their decision-making problems, in particular about the structure of their payoff functions. In the second type of situation, one or more specific users are of interest and knowledge is available or assumptions are made about the structure of the payoff functions of these users. To distinguish between these situations, we refer to the former as the *general-user situation* and the latter as the *specific-user situation*. Moreover, the primary concern in this discussion is the effect of changes in forecast quality (or its aspects) on forecast value, not the actual numerical values of these quantities.

In considering the nature of the quality/value relationship in the general-user situation, it is instructive to compare the general expression for the value of forecasts (VF) with an expression for the mean square error, a measure of accuracy (a specific aspect of quality). Under the assumption that the user's utility function is linear (i.e., utilities and payoffs are linearly related; see Chapters 3 and 4 of this volume), the expression for VF in a single-stage (i.e., static) decision-making problem can be written as follows:

$$\text{VF} = \min_{\alpha} \sum_x t(x)\lambda(\alpha, x) - \sum_f s(f)\min_{\alpha}\sum_x q(x|f)\lambda(\alpha, x), \quad (2.14)$$

where $\lambda(\alpha, x)$ is the loss incurred by the decision maker when action α is taken and (weather or climate) event x occurs. The first term on the right-hand side of equation (2.14) is the expected loss incurred by the user if her decisions are based on the sample climatological probabilities of the events, $t(x)$. This term does not involve the forecasts and can be considered to be a fixed constant [given $t(x)$] for the purposes of this discussion. The second term on the right-hand side of equation (2.14) is the user's expected loss when her decisions are based on the forecasts. Given a particular forecast f, the decision maker is assumed to choose the action that minimizes expected loss for that forecast, with the conditional distributions $q(x|f)$ containing the probabilities required to compute the relevant expected losses. Overall expected loss, as measured by this term, then involves averaging these minimal expected losses over all possible forecasts. When the expected loss associated with climatological probabilities is taken as the zero point on the value scale, the difference between the two terms in equation (2.14) represents the expected value of the forecasts.

The MSE in equation (2.6) can be written as follows:

$$\text{MSE}(f, x) = \sum_f s(f) \sum_x q(x|f)(f - x)^2, \qquad (2.15)$$

since $p(f, x) = q(x|f)s(f)$ (see equation 2.1). Comparison of equation (2.15) and the second term on the right-hand side of equation (2.14) reveals definite similarities between these expressions but also two important differences. First, the second term in VF involves a general loss function $\lambda(\alpha, x)$ (in which the forecast f influences the choice of an action α), whereas the MSE involves a specific "loss" defined in terms of the square of the difference between f and x. Second, the term in VF includes a process of minimization associated with the choice of an optimal action (for each f), whereas the MSE includes no such optimization process. These differences imply (*inter alia*) that the relationship between VF and the MSE is inherently nonlinear (see Chapter 6 of this volume).

The fact that VF and MSE are, in general, different functions of the joint distribution of forecasts and observations has other important implications as well. For example, changes in one or more aspects of quality, as reflected by changes in $q(x|f)$ and/or $s(f)$, generally will affect VF and the MSE (or any other one-dimensional measure of an aspect of quality) differently. As a result, the relationship between VF and MSE will be multivalued rather than single-valued. That is, a range of values of VF will exist for a specific value of the MSE, and vice versa. The existence of a multivalued relationship between VF and MSE implies that decreases (increases) in the MSE can be associated with decreases (increases) in VF. These *accuracy/value reversals* have been investigated by Ehrendorfer and Murphy (1992b) and Murphy and Ehrendorfer (1987), among others. If decisions related to the absolute or relative performance of forecasting systems are based on measures of one or two overall aspects of quality, then they may be subject to such reversals (from the perspective of some users). It is for this reason that we place particular emphasis on the multidimensional structure of verification problems and recommend the use of a suite of verification methods to assess the various aspects of forecast quality.

In specific-user situations, in which the user's payoff function is known or is modeled in terms of a few parameters, the relationship between forecast quality and forecast value can be determined either analytically or numerically. Studies of quality/value relationships in both prototypical and real-world situations have been undertaken in recent years. These studies have confirmed the nonlinear nature of this relationship and have identified other properties (e.g., quality thresholds, convexity) that these relationships appear to possess in many situations of potential interest. Relationships between changes in quality and changes in value in real-world situations are discussed in Chapter 4 of this volume. An in-depth discussion of such relationships in the context of prototypical situations is a primary focus of Chapter 6. These chapters include most if not all of the known references to recent quality/value studies.

7. Conclusion

This chapter has focused on forecast verification, which is characterized here as the process of assessing forecast quality. Forecast quality, in turn, is defined as the totality of the statistical characteristics of the forecasts, the observations, and their relationship embodied in the joint distribution of forecasts and observations. The joint distribution, together with conditional and marginal distributions associated with factorizations of this distribution, constitute the basic elements of a distributions-oriented approach to verification problems. This approach provides insight into the complexity and dimensionality of these problems. Moreover, the various aspects of quality are shown to be directly related to these underlying joint, conditional, and/or marginal distributions. The chapter also describes a method of screening alternative forecasting systems based on the sufficiency relation, as well as several criteria that can be used to screen alternative verification measures.

A basic set of verification methods applicable to all types of weather variables and forecasts has been described and illustrated. These methods include the basic joint, conditional, and marginal distributions themselves, summary measures of these distributions, and various verification measures and terms in decompositions of these measures. The use of these methods in both absolute verification problems and comparative verification problems has been illustrated through applications to verification data samples involving maximum temperature and precipitation probability forecasts. Although maximum temperature forecasts and precipitation probability forecasts differ with respect to both the nature and treatment of the underlying variable and the format of the forecasts, the verification methods applied to the two data sets introduced in Sections 3 and 4 are essentially equivalent. The way in which this body of methods could be tailored to verification problems involving other types of variables and/or forecasts was outlined in Section 5, and this section also contained a brief introduction to some other verification problems and methods. Section 6 discussed the relationship between forecast quality and forecast value, with the objective of providing some insight into the relationship between this chapter and the remaining chapters in the book.

Forecast verification, as described in this chapter, focuses on the use of the verification process to obtain insight into the basic strengths and weaknesses in forecasting performance. This process is an essential component of any effort to assess forecast quality in its full dimensionality or to compare forecasting systems in a rational manner. Since forecast quality is an important determinant of forecast value, detailed assessments of the various aspects of quality are a desirable adjunct to studies of the absolute and/or relative value of weather and climate forecasts.

Acknowledgments

Martin Ehrendorfer provided valuable comments on an earlier version of this chapter. This work was supported in part by the National Science Foundation under Grant SES-9106440.

References

Blackwell, D. (1953). Equivalent comparisons of experiments. *Annals of Mathematical Statistics*, **24**, 265–272.

Blackwell, D. & Girshick, A. (1954). *Theory of Games and Statistical Decisions*. New York: Wiley.

Brier, G.W. (1950). Verification of forecasts expressed in terms of probability. *Monthly Weather Review*, **78**, 1–3.

Brier, G.W. & Allen, R.A. (1951). Verification of weather forecasts. In *Compendium of Meteorology*, ed. T.F. Malone, 841–848. Boston: American Meteorological Society.

Briggs, W.M. & Levine, R.A. (1997). Wavelets and field forecast verification. *Monthly Weather Review*, **125**, in press.

Brooks, H.E. & Doswell, C.A. (1996). A comparison of measures-oriented and distributions-oriented approaches to forecast verification. *Weather and Forecasting*, **11**, 288–303.

Clemen, R.T., Murphy, A.H. & Winkler, R.L. (1995). Screening candidate forecasts: contrasts between choosing and combining. *International Journal of Forecasting*, **11**, 133–146.

Clemen, R.T. & Winkler, R.L. (1987). Calibrating and combining precipitation probability forecasts. In *Probability and Bayesian Statistics*, ed. R. Viertl, 97–110. London: Plenum Press.

DeGroot, M.H. & Fienberg, S.E. (1982). Assessing probability assessors: calibration and refinement. In *Statistical Decision Theory and Related Topics III*, Volume 1, ed. S.S. Gupta & J.O. Berger, 291–314. New York: Academic Press.

Doolittle, M.H. (1885). The verification of predictions. *American Meteorological Journal*, **2**, 327–329.

Ehrendorfer, M. & Murphy, A.H. (1988). Comparative evaluation of weather forecasting systems: sufficiency, quality, and accuracy. *Monthly Weather Review*, **116**, 1757–1770.

Ehrendorfer, M. & Murphy, A.H. (1992a). Evaluation of prototypical climate forecasts: the sufficiency relation. *Journal of Climate*, **5**, 876–887.

Ehrendorfer, M. & Murphy, A.H. (1992b). On the relationship between the quality and value of weather and climate forecasting systems. *Időjárás*, **96**, 187–206.

Epstein, E.S. (1969). A scoring system for probability forecasts of ranked categories. *Journal of Applied Meteorology*, **8**, 985–987.

Finley, J.P. (1884). Tornado prediction. *American Meteorological Journal*, **1**, 85–88.

Gandin, L.S. & Murphy, A.H. (1992). Equitable skill scores for categorical forecasts. *Monthly Weather Review*, **120**, 361–370.

Gilbert, G.K. (1884). Finley's tornado predictions. *American Meteorological Journal*, **1**, 166–172.

Harvey, L.O., Hammond, K.R., Lusk, C.M. & Mross, E.F. (1992). The application of signal detection theory to weather forecasting behavior. *Monthly Weather Review*, **120**, 863–883.

Hoffman, R.N., Liu, Z., Louis, J.-F. & Grassotti, C. (1995). Distortion representation of forecast errors. *Monthly Weather Review*, **123**, 2758–2770.

Hsu, W.-R. & Murphy, A.H. (1986). The attributes diagram: a geometrical framework for assessing the quality of probability forecasts. *International Journal of Forecasting*, **2**, 285–293.

Katz, R.W. & Murphy, A.H. (1990). Quality/value relationships for imperfect weather forecasts in a prototype multistage decision-making model. *Journal of Forecasting*, **9**, 75–86.

Katz, R.W., Murphy, A.H. & Winkler, R.L. (1982). Assessing the value of frost forecasts to orchardists: a dynamic decision-making approach. *Journal of Applied Meteorology*, **21**, 518–531.

Krzysztofowicz, R. (1992). Bayesian correlation score: a utilitarian measure of forecast skill. *Monthly Weather Review*, **120**, 208–219.

Krzysztofowicz, R. & Long, D. (1991a). Forecast sufficiency characteristic: construction and application. *International Journal of Forecasting*, **7**, 39–45.

Krzysztofowicz, R. & Long, D. (1991b). Beta likelihood models of probabilistic forecasts. *International Journal of Forecasting*, **7**, 47–55.

Livezey, R.E. (1995). The evaluation of forecasts. In *Analysis of Climate Variability: Applications of Statistical Techniques*, ed. H. von Storch & A. Navarra, 177–196. New York: Springer-Verlag.

Marschak, J. (1971). Economics of information systems. *Journal of the American Statistical Association*, **66**, 192–219.

Mason, I.B. (1982). A model for assessment of weather forecasts. *Australian Meteorological Magazine*, **30**, 291–303.

Meglis, A.J. (1960). Annotated bibliography on forecast verification. *Meteorological and Geoastrophysical Abstracts and Bibliography*, **11**, 1129–1174.

Mielke, P.W. (1991). The application of multivariate permutation methods based on distance functions in the earth sciences. *Earth–Science Reviews*, **31**, 55–71.

Murphy, A.H. (1971). A note on the ranked probability score. *Journal of Applied Meteorology*, **10**, 155–156.

Murphy, A.H. (1973). Hedging and skill scores for probability forecasts. *Journal of Applied Meteorology*, **12**, 215–223.

Murphy, A.H. (1988). Skill scores based on the mean square error and their relationships to the correlation coefficient. *Monthly Weather Review*, **116**, 2417–2424.

Murphy, A.H. (1991). Forecast verification: its complexity and dimensionality. *Monthly Weather Review*, **119**, 1590-1601.

Murphy, A.H. (1993). What is a good forecast? An essay on the nature of goodness in weather forecasting. *Weather and Forecasting*, **8**, 281–293.

Murphy, A.H. (1996). The Finley affair: a signal event in the history of forecast verification. *Weather and Forecasting*, 11, 3–20.

Murphy, A.H., Brown, B.G. & Chen, Y.-S. (1989). Diagnostic verification of temperature forecasts. *Weather and Forecasting*, 4, 485–501.

Murphy, A.H. & Daan, H. (1985). Forecast evaluation. In *Probability, Statistics, and Decision Making in the Atmospheric Sciences*, ed. A.H. Murphy & R.W. Katz, 379–437. Boulder, CO: Westview Press.

Murphy, A.H. & Ehrendorfer, M. (1987). On the relationship between the accuracy and value of forecasts in the cost–loss ratio situation. *Weather and Forecasting*, 2, 243–251.

Murphy, A.H. & Epstein, E.S. (1989). Skill scores and correlation coefficients in model verification. *Monthly Weather Review*, 117, 572–581.

Murphy, A.H. & Wilks, D.S. (1996). Statistical models in forecast verification: a case study of precipitation probability forecasts. *Preprints, Thirteenth Conference on Probability and Statistics in Atmospheric Sciences*, 218–223. Boston: American Meteorological Society.

Murphy, A.H. & Winkler, R.L. (1987). A general framework for forecast verification. *Monthly Weather Review*, 115, 1330–1338.

Murphy, A.H. & Winkler, R.L. (1992). Diagnostic verification of probability forecasts. *International Journal of Forecasting*, 7, 435–455.

Murphy, A.H. & Ye, Q. (1990). Comparison of objective and subjective precipitation probability forecasts: the sufficiency relation. *Monthly Weather Review*, 118, 1783–1792.

Peirce, C.S. (1884). The numerical measure of the success of predictions. *Science*, 4, 453–454.

Potts, J.M., Folland, C.K, Jolliffe, I.T. & Sexton, D. (1996). Revised "LEPS" scores for assessing climate model simulations and long-range forecasts. *Journal of Climate*, 9, 34–53.

Stanski, H.R., Wilson, L.J. & Burrows, W.R. (1989). Survey of common verification methods in meteorology. Research Report No. 89-5, 114 pp. Toronto: Atmospheric Environment Service.

Swets, J.A. (1988). Measuring the accuracy of diagnostic systems. *Science*, 240, 1285–1293.

Wilks, D.S. (1995). *Statistical Methods in the Atmospheric Sciences*. New York: Academic Press.

Winkler, R.L. & Murphy, A.H. (1968). "Good" probability assessors. *Journal of Applied Meteorology*, 7, 751–758.

3

The value of weather information

STANLEY R. JOHNSON
and
MATTHEW T. HOLT

1. Introduction

The existing system for sensing, recording, and reporting weather conditions and producing forecasts has been developed mainly in response to demands of specific clients. Weather conditions and forecasts provided for airline navigation and agricultural production management are but two examples. The result is that the system for producing, storing, and disseminating weather data and forecasts has strong historical linkages to the demands of major clients and the sensing and recording technologies available at the time of implementation. Location of first-order stations at airports, the cooperative observer system, frequency of reporting, and the levels in the atmosphere at which data are observed all can be viewed as having a user-based history.

With the advent of new sensing, recording, and reporting technologies, and changing needs of existing clients and the entry of new clients, there has been a growing effort to justify economically the system supplying these services. Weather data and forecasts are produced to a large extent by the public sector and made available at a highly subsidized user cost; that is, the data are public goods. To provide an economic rationalization for the production and dissemination system, it must be shown that the rate of return, or benefit relative to cost, is consistent with that available from alternative employment of societal resources. For this calculation, the relevant cost and benefit concepts are, of course, social, as opposed to private or individual.

The effort to justify economically the weather information system has resulted in a number of research activities and suggested organizational changes. One way to value the weather information system would be to make it private in some way and then estimate its value from sales of privatized information. Proponents agree that this alternative would produce the service in such a manner that the market would automatically value it. This approach to valuation and resource allocation has recently received increased attention as indicated by the initiation of user charges for certain types of weather information and related services. Although still at an early stage, this method of organizing the production and delivery system for weather information can be viewed as "testing" the market for these services. In fact, extensive privatization of weather information services has already been implemented in countries such as New Zealand and Sweden.

It is, however, important to recognize that a market-based approach may not be efficient for valuing weather information or allocating resources. Weather information is a nonrival and, to a large extent, nonexcludable commodity. That is, two or more consumers can simultaneously use the same unit of weather information (nonrival) and it is not, in general, possible to prevent certain groups or individuals from using available weather information (nonexcludable). The implication is that market equilibrium is not optimal, since the economic externalities of weather information are not incorporated into individual decision making. Theoretically, either an artificial market must be established (i.e., a system that artificially assigns property rights for weather information) or a socially optimal tax-subsidy scheme must be implemented, if an efficient resource allocation to the weather information system is to be attained (Malinvaud, 1971). Hence information obtained from testing the market must be viewed cautiously as an input to the design of a socially optimal weather information service.

To complement the market experiments, research has been undertaken to develop more formal valuations of weather information systems in specific contexts. This research has involved one or two primary emphases. The first emphasis is related to determining the value of weather services, or specific components of those services, for both individual decision-making units and society as a whole. This applied research is carried out with the goal of actually assigning monetary values to the components of the weather information studied. Applied studies of information value at the individual decision level are the most numerous and the most cogent. The second research emphasis is on developing appropriate methods for measuring and estimating individual and societal values of weather information services.

The present chapter has the objective of reviewing the progress that has been made in valuing the weather information system. First, selected concepts from the economics of information are reviewed. The intent is to provide a general framework for analyzing valuation methods currently employed. Then, selected studies that have estimated the economic value of particular components of the weather information system are discussed. This exercise — comparing valuation theory with applications for the weather information system — raises a number of questions. Important issues posed by the questions pertaining to design of valuation studies, privatization, and resource allocation are then examined. Finally, a few observations are provided on the progress in valuation methods and the potential for new analyses to improve the basis for designing and organizing the weather information system.

A final caveat is in order. In this chapter we do not discuss what we refer to as "impact assessment" studies, which seek to determine the impact of a weather-related event on a particular segment of the economy or society. An excellent example of these types of studies is provided by Roll (1984), who examined the causal relationships between temperature and rainfall near Orlando, Florida, and the price of frozen concentrated orange juice contracts. While impact assessment studies provide valuable information for policy makers and for decision makers about the isolated effects of weather-related events on specific markets or market participants, these studies do not in any way represent true value-of-information assessments in the context of the system for producing and disseminating weather data and forecasts.

2. Economics and the value of information

The economic theory of information value has progressed significantly in recent years. With the development of the von Neumann–Morgenstern utility hypothesis, and the refinement of decision theory under uncertainty (Arrow, 1965; Pratt, 1964), the integration of value-of-information theory into mainstream economic thinking has occurred rapidly. In short, the development of risk or uncertainty theory has provided a basis for reconciling a number of important issues related to the value of information in society, investment in the production of information, impacts on price determination, relationships between information and prices, etc. (Fama, 1970; Grossman and Stiglitz, 1976; Gould, 1974; Hayek, 1945; Hess, 1982; Hirshleifer, 1971, 1989; Kunkel, 1982; Marschak, 1971; McCall, 1965; Riley, 1979). Surveys of value-of-information theory that provide an integration and synthesis of available results can be found in Hirshleifer and Riley (1979, 1992) and Stigler (1961). In addition, there have been a number of studies in the management science/statistics literature that have advanced basic concepts and methods for information valuation (e.g., Blackwell, 1953; Blackwell and Girshick, 1954; DeGroot, 1970; Hilton, 1981; Raiffa, 1968; Winkler, 1972). The present discussion reviews key theoretical concepts that are helpful in interpreting available empirical results on the valuation of weather information. We also suggest a possible framework for improving the generality and scope of future investigations.

2.1. Information and individual valuation

The approach in modern economic theory is simply to view information as a factor in the decision process that can be used by economic agents (or decision makers) to reduce uncertainty (i.e., so-called Bayesian decision theory). A stylized individual decision model illustrating the central concepts of the theory can be developed as follows. Subjective probabilities, along with the assumption that agents can assign a unique utility ranking to all possible outcomes, are keys to the theory (von Neumann and Morgenstern, 1944; Winkler, 1972).

Consider a set of actions or "terminal" moves, $a = 1, \ldots, N$, that an economic agent (i.e., a user of weather information) can choose among, and a set of possible states of the world (i.e., weather events in this case), $s = 1, \ldots, M$, over which the agent is assumed to have no direct control. A finite number of actions a, as well as a finite number of states s, are assumed for convenience; it is, however, a straightforward matter to extend the model to accommodate an infinite number of actions and states. The consequences resulting from each possible action and each possible state of the world $c(a, s)$ are, in all instances, presumed known to the agent. Furthermore, the agent is assumed to have the ability to rank the possible consequences of each action according to relative desirability. That is, the individual is assumed to possess a utility function u whose composition with c determines a preference relation $u[c(a, s)]$, which is defined over the set of all possible actions and consequences. In the present framework, uncertainty arises because the agent must choose an action prior to observing the realized state. This uncertainty is characterized formally by a set of subjective probabilities associated with realizing each of the various states.

The agent is assumed to have a prior probability distribution on the possible (finite) states of the world, with the subjective probabilities for the agent being denoted by p_s. In practice, these subjective probabilities may be based on historical weather data and are termed "climatological information" elsewhere in this volume. The individual decision problem is then to select the action a_o that satisfies

$$E\{u[c(a_o, \cdot)]\} = \max_a E\{u[c(a, \cdot)]\} = \max_a \left\{\sum_s p_s u[c(a, s)]\right\}, \quad (3.1)$$

where $E\{u[c(a, \cdot)]\}$ (the dot indicates expectation with respect to the distribution p_s for the state s) is the expected utility to the individual decision maker if action a is taken. Note that if the decision maker is risk neutral, the utility maximization problem is equivalent to choosing the action that maximizes the expected profit. This follows from the fact that marginal utility is constant in the risk neutral case, irrespective of the level of income.

The agent's subjective probability distribution on states of the world can be modified by acquiring information. In the present context, information can be viewed as a set of possible messages. These messages, denoted $i = 1, \ldots, I$, provide the basis for revising the probabilities associated with each state of the world. This revision process may, in turn, lead to a different choice of "terminal" action. The decision maker, however, does not know in advance which among the possible set of messages will be received. This result implies that *ex ante*, or before the message is received, the decision maker must determine a subjective probability q_i of receiving message i. The probability q_i is in turn related to the conditional probabilities or likelihoods $q_{i,s}$ of receiving message i given state s, and is determined by

$$q_i = \sum_s q_{i,s} p_s, \quad (3.2)$$

where p_s represents the previously defined subjective (or prior) probability associated with state s.

Bayes' theorem, in combination with the message probabilities defined in equation (3.2), provides a basis for revising or updating the probabilities attached to each state of the world. More specifically, after receiving message i, the decision maker can determine the posterior probability of state s given message i by

$$p_{s,i}^* = \frac{q_{i,s}p_s}{q_i}, \tag{3.3}$$

where $p_{s,i}^*$ denotes the posterior probability. We note that, in practice, the weather forecasting system can be viewed as directly producing these posterior probabilities (instead of the likelihoods).

Figure 3.1 provides an illustrative example of a Bayesian revision for the case of continuous s. (Here a continuous state space is assumed simply to facilitate graphical representation.) The likelihood function shows the probability of receiving message i given state s (i.e., $q_{i,s}$). The prior probability distribution defines unconditional probabilities p_s associated with state s before message i has been received. Initially the individual believes relatively higher values of s are more likely, as indicated by the location of the prior distribution. The likelihood function, however, shows that the probability of receiving message i is higher for relatively low values of s. The revised or posterior probability distribution then represents a composite of the prior distribution and the likelihood function, as determined by Bayes' theorem in equation (3.3). As is clear from equation (3.3), the more certain are the prior beliefs p_s, the more closely the posterior distribution will resemble the prior distribution irrespective of the values of q_i and $q_{i,s}$. This phenomenon is referred to in value-of-information theory literature as the distinction between "hard" versus "soft" prior beliefs about possible states of the world (Hirshleifer and Riley, 1992). Moreover, the foregoing observation suggests that the greater the level of initial confidence pertaining to a particular state s, the lower the value the individual will attach to receiving message i (but this is not necessarily the case in general). In weather applications, prior information is typically very diffuse.

Values of additional or new information (i.e., the message) are based on the expected utility from the more informed decision as

Probability Density

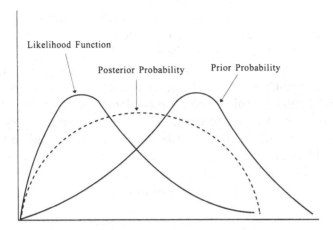

Figure 3.1. Bayesian probability revision.

compared to the expected utility without the information. Several measures of the economic value of imperfect information exist (Hilton, 1981). One measure of the value of information is the difference in expected utility,

$$V_i = E_i\{u[c(a_i, \cdot)]\} - E\{u[c(a_o, \cdot)]\}, \qquad (3.4)$$

where E_i denotes expectation with respect to the posterior probabilities. The information decision problem for the individual decision maker is then one of comparing the two optimal choices, a_o and a_i. Under the first optimal choice, a_o, which is the optimal action given only prior information, the information or message i has yet to be received. Under the second optimal choice, a_i, where a_i denotes the optimal action given message i, the information has been received and processed in accordance with equation (3.3) to form $p^*_{s,i}$, the posterior probability associated with state s. Alternatively, another valuation measure is the "demand value," that is, the maximum amount that the decision maker would be willing to exchange for the system (Hilton, 1981).

The above illustrative valuation problem is *ex post* in nature in the sense that information is valued *after* the message has been received, but before the actual state of the world is known. While this simplified version of the information decision problem is useful and, in fact, characterizes much empirical work on information valuation, at the individual level, it must be stressed that the problem is not yet formulated as typically perceived by the decision maker. The information decision as typically encountered is an *ex ante* problem, in that the decision maker does not know *a priori* which message will ultimately be received.

To formulate the *ex ante* information valuation problem, consider the agent's decision about additional or new information. The individual in this circumstance does not know in advance which message will be received. To simplify, assume the agent purchases information from a vendor. Alternatively, this information could be provided by a public agency (e.g., a national weather service). If the information has no cost, then the *ex ante* value of information to the individual decision maker is characterized by

$$V^* = \sum_i q_i E_i\{u[c(a_i, \cdot)]\} - E\{u[c(a_o, \cdot)]\}, \qquad (3.5)$$

where the probability distribution q_i reflects the agent's uncertainty about receiving message i. The decision problem in this extended context is thus one of summing or integrating over all possible messages and associated probabilities. The *ex ante* valuation implied by equation (3.5) represents the expected utility gain associated with receiving information. As such, equation (3.5) is not a monetary measure of information value, but rather a utility-based measure. As already mentioned in conjunction with equation (3.4), an alternative to equation (3.5) is the demand value of this information. This measure is equivalent to equation (3.5) if the utility function is linear or negative exponential.

2.2. Information and market valuation

In principle the value-of-information problem for the individual agent can be developed in a straightforward manner; the market determination of information value, however, is more difficult. Two issues complicate the market valuation problem: (i) the equilibrium condition for the market and how this equilibrium condition is modified by the introduction of additional information; and (ii) the aggregation of individual responses to produce market level supply and/or demand functions used in establishing economic value. Of course, the *ex ante–ex post* decision problem remains (Choi and Johnson, 1987).

Recent developments in rational expectations theory provide a tractable way of formulating economic models when agents must make decisions on the basis of imperfect information. The rational expectations hypothesis essentially implies that individuals understand the underlying structure or fundamentals of the market in which they participate and act on that information (Muth, 1961). Although there are other expectations theories, and while the informational assumptions of the rational expectations hypothesis are rather rigid, the rational expectations approach is nevertheless useful as a benchmark against which to compare results of other expectations hypotheses. Moreover, the rational expectations approach is attractive for its consistency with other behavioral assumptions included in the economic model specification (i.e., expected utility maximization, etc.).

Results that are based on formulating economic models with rational expectations and developing appropriate microeconomic foundations for market equilibria when market participants face uncertainty have been obtained only recently (Newberry and Stiglitz, 1981; Wright, 1979; Innes, 1990). Many studies that have applied the rational expectations framework have evaluated price or revenue stabilization policies in competitive markets. They show that the benefits of intervention in competitive markets result from the more stable environment provided for producers and/or consumers. A modest extension of these results is to consider market intervention through added information about uncertain events, even if at a cost to participants (Babcock, 1990).

As a simple illustration of the above concepts, consider a case in which N identical producers of a homogeneous good face a demand schedule with a stochastic component. Producers are presumed to make production decisions *ex ante*, before the output demand is realized. The number of market participants, N, is also assumed large enough that the industry can be considered competitive. If q represents output of a single representative producer, then $Q = Nq$ is industry output. A general representation of the stochastic market demand function is given by

$$P = P(Q,\mu), \ \partial P/\partial Q < 0, \tag{3.6}$$

where μ is a random variable with distribution function $G(\mu)$. For a given level of market output Q, the distribution function

$G(\mu)$ determines completely a price distribution $F(P|Q)$, and an expected inverse demand function

$$\overline{P}(Q) = \int_0^\infty P dF(\Gamma|Q) - \int_{-\infty}^\infty P(Q, \mu) dG(\mu) \qquad (3.7)$$

In previous studies the risk averse producer has been assumed to have a subjective distribution $F^e(P)$ of possible price outcomes (Baron, 1970; Sandmo, 1971; Leland, 1972). By invoking the rational expectations hypothesis in the present model, it is assumed that the true distribution $G(\mu)$ ultimately coincides with producers' subjective beliefs. That is, once producers make a subjective estimate Q^e of industry output Q, the subjective distribution $F^e(P) = F(P|Q^e)$ is completely determined.

Given the subjective price distribution $F^e(P)$, a producer will maximize expected utility of profit

$$E[U(\pi)] = \int_0^\infty U[Pq - c(q)] dF^e(P), \qquad (3.8)$$

where $U(\cdot)$ is a von Neumann–Morgenstern utility function and $c(q)$ is an appropriate cost function. Since the representative producer's optimal output q^* depends on the subjective estimate Q^e, we can write optimal output as $q^* = q^*(Q^e)$. Thus, industry output is expressed as

$$Q^* = Nq^*(Q^e) = H(Q^e), \qquad (3.9)$$

where $H(\cdot)$ denotes the mapping between expected output, Q^e, and optimal industry production, Q^*. The rational expectations hypothesis, as used in this simplified context, implies that firms' subjective price distribution $F^e(P)$ will equal the actual price distribution $F(P)$ (Choi and Johnson, 1991). Of course, at the end of the period when market demand is realized, producers only observe the actual market price and not a distribution. An important feature of the present model is that firms can verify *ex post* the rationality of their production decisions by comparing Q^e with Q. Because $F(P|Q^e) = F(P|Q)$ if and only if $Q = Q^e$, the rational expectations hypothesis implies that market equilibrium occurs in the present context only when actual and anticipated output are equal.

The above framework can be used to investigate the market valuation of information. As Chavas and Johnson (1983), Pesaran (1987), and others point out, a number of interesting parallels exist between value-of-information theory and the formation of rational expectations. The exact nature of the linkage depends on the amount of information available and its cost at the time the firm makes production decisions. With additional information, producers could revise their subjective estimates of the price distribution $F^e(P)$, as well as their subjective estimates of all relevant distribution parameters. Additional information should, in general, improve resource allocation and enhance market efficiency. Even if additional information did nothing more than bring about a change in dispersion without changing the distribution's mean (Rothschild and Stiglitz, 1970), however, we would still expect different outcomes or market results. The reason for this is that any change in the dispersion parameter of $F^e(P)$, even if the centrality parameter remains unchanged, will affect optimal output decisions for risk averse producers that maximize expected utility.

A number of refinements and extensions can be made to the above model. The simplified model is included only to suggest the complexity of the market valuation problem. For instance, the model could be extended to include a stochastic production process. In this case, q becomes a random variable with a distribution conditioned on the level of inputs. Producers would then determine subjective estimates of the joint distribution of P and Q in a manner consistent with the rational expectations approach. Not only does the rational expectations framework have important implications for estimating the market value of information, but it also raises a number of questions about the "appropriateness" of conventional valuation theory under a scenario of uncertainty, as well as questions about the empirical methods presently employed in estimating market relationships from *ex post* or observed market outcomes for use in valuation exercises.

It is now recognized that standard producer surplus is an inappropriate measure of welfare under conditions of uncertainty (see, e.g., Pope, Chavas, and Just, 1983). By employing the rational expectations market equilibrium framework, the information valuation problem can be addressed more systematically. The *ex post–ex ante* problem remains, however. Most surplus measures used in the economic theory of value are *ex post*, and based on the assumption that all relevant economic variables are known with certainty (i.e., are nonstochastic). Recently it has been shown that these concepts must be modified if the information valuation problem is viewed in an *ex ante* context (Choi and Johnson, 1987).

More specifically, three commonly applied welfare measures are Marshallian consumers' surplus, compensating variation, and equivalent variation. Marshallian consumers' surplus is simply the area under the demand curve and will be explained more fully in Section 3.2. Compensating variation is defined as the additional income necessary, after a price change, to restore an individual to the original level of well-being that was enjoyed before the price change occurred, assuming the new price level holds. Equivalent variation is the amount of income necessary, after a price change, to restore the individual to the original level of utility, assuming that the initial price still holds. The distinction is that compensating variation uses an "after-price change" base, while equivalent variation uses a "before-price change" base. Willig (1976) has illustrated under very general conditions that Marshallian consumers' surplus closely approximates compensating variation. This fact, coupled with ease of application, has resulted in the continued use of Marshallian surplus measures in empirical valuation studies (a classic reference is Hayami and Peterson, 1972).

An additional complication arises, however, if price is a random variable. The most common approach in this instance is to recognize that the surplus measures are also random variables and that their expectations will provide an indication of average benefits accruing to an individual (Waugh, 1944). It has been shown recently, though, that these concepts must be modified if the information valuation problem is to be viewed *ex ante*. In particular, it has been demonstrated that expected Marshallian consumers' surplus (and, consequently, expected compensating variation) is a valid welfare measure only in the special case when marginal utility of income does not depend on price (Turnovsky, Shalit, and Schmitz, 1980; Rogerson, 1980). Even with these clear conceptual problems, expected Marshallian consumers' surplus is widely used (e.g., Burt, Koo, and Dudley, 1980; Taylor and Talpaz, 1979).

Several authors have proposed measures to determine correctly the consumer benefits in a stochastic setting. Specifically, Anderson (1979a) and Helms (1985) have argued that *ex ante* compensating and equivalent variations are improved measures of consumer benefits when price is a random variable. These *ex ante* measures are appropriate in the present context for evaluating weather information. The relevant compensation experiment is to determine how much income the potential user would be willing to forgo in exchange for the information service before the outcome is known.

In previous value-of-information studies, expectations of Marshallian consumers' surplus measures have typically been applied to assess the economic benefits of information (i.e., Bradford and Kelajian, 1977). As indicated in the preceding discussion, however, Marshallian consumers' surplus measures are inappropriate if the valuation problem involves stochastic prices, and is thus *ex ante*. Furthermore, because expected compensating variation employs an after-price change base, it is the amount of monetary income necessary, on average, to compensate a consumer for facing prices in a no-information environment if the compensation is paid *after* the random price is observed. Clearly, this measure does not reflect the willingness of the individual to pay for an information service before observing the actual price outcome.

Although *ex ante* compensating and equivalent variations are appropriate welfare measures in a stochastic price setting, these measures are of limited practical value. They require information about risk attitudes of consumers and about properties of the direct utility function that are difficult to obtain. These limitations may explain the continued use of expected consumer's surplus measures in empirical valuation studies. Alternatively, Hausman (1981) has shown that compensating and equivalent variation measures — and consequently the expected values of these measures — can be recovered from many common forms of estimable demand functions (e.g., linear, double logarithmic, etc.). Choi and Johnson (1987) have provided further justification for the use of expected equivalent variation in empirical applications. They show that expected equivalent variation and *ex ante* equivalent variation are identical if the individual is risk neutral. More important, they demonstrate that expected equivalent variation provides a lower bound for *ex ante* equivalent variation when individuals are risk averse in income. These favorable aspects of expected equivalent variation suggest it will be more widely applied in future studies of information value.

3. Review of selected value-of-information studies

In recent years, a number of studies have attempted to value the weather information system, as well as particular weather information collection and dissemination systems. Studies examining these and related issues are too numerous to allow a complete enumeration here. Our objective is to highlight briefly selected studies illustrative of the kinds of issues examined and the methods used to value weather information. In addition, emphasis is placed on the studies that have provided actual value estimates of improved weather information. As already indicated, the literature on the value of weather information generally falls into two broad categories: (i) the value of weather information to individual decision makers; and (ii) the value of weather information at the market level. Examples of studies conducted under both of these categories are presented here; more examples and further details on studies under category (i) can be found in Chapter 4 of this volume.

3.1. Individual decision applications

Of the two areas, primary research emphasis has been placed on valuing weather information at the individual decision-making level. As discussed in Section 2.1, individual valuation studies are generally couched in a decision-analytic framework. Many of these studies involve a "cost–loss situation," in which a decision maker must choose one of two actions: protect an activity or operation at a known cost or face the risk of, perhaps catastrophic, loss. Upon receipt of the forecast information, initial probabilities for each state of nature can then be revised in accordance with Bayes' theorem. See Section 2.1 and, specifically, Bayes' theorem as presented in equation (3.3), for additional details. In all cases, the decision maker is assumed to choose the action maximizing expected return (minimizing expected expense) or maximizing expected utility.

Table 3.1 summarizes a number of applied studies that have examined the value of weather information for individual decision-making units. Studies that have used a Bayesian framework for analyzing the value of weather information include those by Baquet, Halter, and Conklin (1976), Katz, Murphy, and Winkler

(1982), Stewart, Katz, and Murphy (1984), Mjelde et al. (1988), Hashemi and Decker (1972), and Byerlee and Anderson (1969). Note that the Baquet et al. study relied on the more general demand value measure of the value of information mentioned earlier. Other Bayesian decision-analytic studies are reviewed in Chapter 4 of this volume. Anderson (1979b), Lave (1963), and Sonka, Changnon, and Hofing (1988) have used a "cost–loss" approach in valuing weather information. Additional studies have determined the value of weather information to individual decision-making units by using less structured subjective measures. These studies typically involve surveys of users in which the respondents are asked to estimate subjectively the value of an information service. One example is the survey by Ewalt, Wiersma, and Miller (1973). See also Chapter 5 of this volume for a more detailed discussion of the survey approach.

3.2. Market applications

While studies investigating information value to individuals are numerous, far fewer inquiries into the social value of weather information have been made. Most studies assessing the social value of information use *ex post* Marshallian surplus or benefits measures. It is assumed that producers and/or consumers make economic decisions with uncertainty about the possible outcomes. The benefits of information result either because markets have a temporal dimension (i.e., inventory levels are adjusted) or because economic agents have flexibility to adjust to new (more current) information.

Figure 3.2 illustrates how the Marshallian framework is typically applied to estimate the social returns from improved weather information when markets are temporally linked. Holders of storage (i.e., arbitragers) decide the level of inventories to carry forward from period 1 on the basis of expectations about production in the second period. If it were known that quantity Q_1 would be produced in the first period and quantity Q_2 in the second period, then social value would be maximized by choosing an inventory level equating prices in the two periods (in the absence of storage costs and time preference for money). That is, with identical linear demands, the optimal inventory divides total output $(Q_1 + Q_2)$ equally between the two periods. Now, suppose agents do not know Q_2 — perhaps it depends on stochastic climate conditions

Table 3.1. Summary of selected value-of-information studies: individual decision applications

Investigators	Subject	Weather Characteristics	Weather Impact Variables	Information Concept	Value System	Valuation Method	Conclusion
Baquet et al. (1976)	Value of frost fore. to pear orchard managers	Temp.	Bud damage & yield loss	Daily min. temp. fore., hist. wea. info.	Individual	Bayesian, expected utility max.	US NWS frost fore. had approx. value of $13.32/ha-day for risk averse decision makers.
Katz et al. (1982)	Value of frost fore. to orchardists in Yakima Valley	Temp.	Bud damage & yield loss	Daily min. temp. fore., hist. wea. info.	Individual	Bayesian, expected cost min., Markov decision process	Value ranged from $667 to $1,997/ha-yr.
Stewart et al. (1984)	Value of frost fore. to apple orchardists in Yakima Valley	Temp., dew pt.	Bud damage & yield loss	Daily min. temp. fore. & post fore. temp. & dew pt.	Individual	Bayesian, expected cost min.	Approx. $1,885/ha-yr.
Wilks & Murphy (1985)	Value of prec. fore. for haying/pasturing decisions in west. Oregon	Prec., temp.	Net income from pasturing or haymaking; hay quality	Seasonal prec. fore.	Individual	Bayesian, expected utility max.	Value of current fore. ranges from $0.00 to $1.41/ha-day.
Hashemi & Decker (1972)	Irrigation scheduling in corn prod.	Prec.	Irrigation freq.	Prec. prob. fore.	Individual	Bayesian	Reduction in magnitude & freq. of supplemental irrigation.

Reference							
Ewalt et al. (1973)	Value of prec. & field condition fore. to Indiana farmers	Prec., field conditions	None	Prec. & field condition fore.	Individual	Survey (subjective)	Value of fore. positively related to soil types. Highest values obtained for spring mos.
Lave (1963)	Value of better wea. info. to raisin industry	Prec., deg. days	Grape yields & uses (i.e., raisins, crushing, etc.)	Perfect prec. fore.	Individual, market	Cost–loss & impact on industry profits	Value of perfect 3-week fore. is $225/ha. Partial equilibrium analysis shows industry profits fall with improved wea. fore.
McQuigg & Thompson (1966)	Natural gas demand in Columbia, MO	Heating deg. days	Demand for natural gas	Fore. of heating deg. days	Individual	Loss function	Value of wea. info. depends on ability of user to translate effectively such info. into economic terms.
Tice & Clouser (1982)	Value of wea. info. to individual corn producers	Min. & max. soil & air temp. & daily prec.	Corn & soybean yields	Prob. knowledge of wea. indices	Individual	Accounting, expected profit max.	Utilizing current wea. info. & prob. of future wea. events results in increased returns of $3.66 to $9.86/ha.
Anderson (1979b)	Value of extended-period wea. fore. in pea prod. & logging	Temp., prec.	Ripening date for peas. Road improvement costs for logging	Prob. knowledge of wea. data	Individual	Cost–loss	Savings level varies depending on conditional prob. of "bad" outcome given that unfavorable outcome has been predicted.

Table 3.1 continued.

Investigators	Subject	Weather Characteristics	Weather Impact Variables	Information Concept	Value System	Valuation Method	Conclusion
Byerlee & Anderson (1969)	Value of prec. predictors in wheat yield response functions	Prec.	Wheat yield through nitrogen, prec., soil moisture interactions	Prediction of annual prec. trends	Individual (*ex ante*)	Bayesian, expected profit max.	Values of prec. predictors ranged from 0.7 to 89.0 cents/ha.
Brown et al. (1986)	Value of seasonal wea. fore. to wheat farmers in Great Plains	Prec., soil moisture	Wheat yields	Clim. info. & imperfect fore. of wea. conditions	Individual	Expected profit max.	Value of current & perfect fore. estimated between $0.00 & $196.62/ha, resp.
Sonka et al. (1987)	Value of seasonal wea. fore. to Illinois corn producers	Temp., prec., solar radiation	Corn yields	Hist. freq. of clim. conditions	Individual	Expected profit max.	Value of perfect annual fore. varies from $21.20 to $45.99/ha-yr. Fore. for early summer crop stage were most valuable.
Bosch & Eidman (1987)	Value of soil moisture & wea. info. to irrigators	Temp., prec., solar radiation	Irrigation schedule	Perfect fore. of prec. & evapotranspiration for 3-day horizons	Individual	Expected profit, generalization stochastic dominance	Value of perfect wea. & soil info. varies from $2.47 to $40.28/ha, depending on degree of risk aversion.

Mjelde & Cochran (1988)	Bounds on wea. info. value to corn producers	Temp., prec., solar radiation, wind	Corn yields	Perfect fore. of seasonal clim. indices	Individual	Net returns, stochastic dominance	Value of perfect fore. varies between $0.00 & $218.00/ha-yr, depending on prior knowledge & deg. of risk aversion.
Mjelde et al. (1988)	Value of seasonal wea. fore. to corn producers	Temp., prec., solar radiation, wind	Corn yields	Seasonal wea. fore. at various crop stages	Individual	Bayesian, expected net returns max.	Expected value of perfect fore. varies from $17.95 to $28.46/ha-yr, for different levels of prior knowledge.
Sonka et al. (1988)	Value of wea. fore. to major seed corn producing firm in Midwest	Temp., prec.	Corn yields	Seasonal temp. & prec. fore.	Individual	Cost–loss	Savings from perfect info. between 2 & 5% of production cost. Fore. of adverse conditions, accurate only 50% of time, had 2/3 value of perfect fore.
McGuckin et al. (1992)	Value of field clim. info. to irrigators	Temp., prec., soil moisture	Corn yields	Soil moisture as function of temp., prec., & soil type	Individual	Av. cost reduction	Moisture sensors can improve technical efficiency by 3.9% on av. Value of info. depends on producer's technical efficiency, & ranges from $40 to $58/ha.

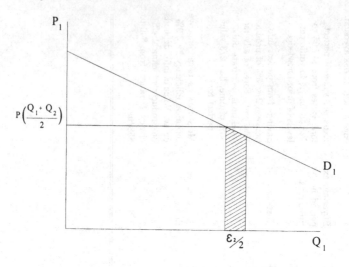

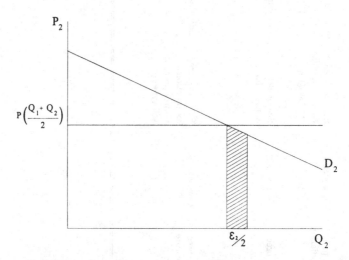

Figure 3.2. Welfare effect of forecast error: (top) period 1 consumption; (bottom) period 2 consumption.

— and instead use $Q_{2,1}$, which differs from Q_2 by a forecast error ε_2. Arbitragers will then hold in inventory an amount equal to $(Q_1 - Q_{2,1} - \varepsilon_2)/2$, which is $\varepsilon_2/2$ less than the amount $(Q_1 - Q_2)/2$. That is, if ε_2 is positive, too little inventory is held relative to a situation of perfect foresight.

The dollar loss in consumption in period 2 is given by the hatched trapezoid in Figure 3.2 (bottom). This area shows the extra consumption value attached to perfect foresight. Period 1 consumption is correspondingly larger than it otherwise would have been, with the resulting dollar gain due to the additional consumption equal to the hatched area in Figure 3.2 (top). Since demand is nonrandom, the net loss in value relative to what would have occurred with a perfect forecast is given by the area of the rectangle with base $\varepsilon_2/2$ and height $\beta\varepsilon_2/2$, where $-\beta$ is the slope of the (linear) inverse demand function. Thus, the net social loss is $\beta\varepsilon_2^2/4$. An analogous argument holds when ε_2 is negative. Assuming that $E(\varepsilon_2) = 0$, the expected value of the social loss is given by $\beta\sigma^2/4$, where σ^2 represents the forecast error variance.

Increasing the accuracy of the forecast (i.e., reducing the forecast error variance) can be evaluated to determine the marginal social benefit of improvements in forecasting accuracy. A similar framework can be used to evaluate the social benefit of improved intrasoasonal forecasts when producers have the flexibility to adjust production responses to the new information. Various forms of the above approach have been used to value information for crop and livestock forecasting systems; for example, Hayami and Peterson (1972), Bradford and Kelajian (1977, 1978), Freebairn (1976), and more recently Adams et al. (1995) (see Table 3.2).

Only a few studies have attempted to estimate directly the social value of weather forecasting per se. Furthermore, the only studies conducted to date in a rational expectations framework are by Antonovitz and Roe (1984) and Babcock (1990), although in neither case did the analysis focus on public policy aspects of improved forecast performance. Lave (1963) examined the potential worth of improved forecasts for the raisin industry, but his conclusion was limited to the argument that better forecasts would result in larger output and, because of the inelastic nature of raisin demand, prices and revenues would fall. He did not, for example, directly estimate consumer benefit. One recent attempt to assess the economic returns from improved weather forecasts was undertaken by Adams et al. (1995). Their study focused on estimating the value of improved El Niño–Southern Oscillation forecasts to the agricultural sector in the southeastern United States. The results indicated that, under a free-market setting (i.e., in the absence of farm programs), the value of perfect forecasts is $145 million (in 1990 dollars), while the value of imperfect but improved forecasts is $96 million (again, in 1990 dollars).

Table 3.2. Summary of selected value-of-information studies: market applications

Investigators	Subject	Weather Characteristics	Weather Impact Variables	Information Concept	Value System	Valuation Method	Conclusion
Lave (1963)	Value of better wea. info. to raisin industry	Prec., deg. days	Grape yields & uses (i.e., raisins, crushing, etc.)	Perfect prec. fore.	Individual, market	Cost-loss & impact on industry profits	Value of perfect 3-week fore. is $225/ha. Partial equilibrium analysis shows industry profits fall with improved wea. fore.
Freebairn (1976)	Value of commodity price outlook info.	—	Supply response is function of producer price fore. in previous periods	Improvement in accuracy of commodity price fore.	Market	Marshallian surplus	Potential gross benefits for wool, lamb, wheat, barley, & potato markets in Australia was at least 1% of gross value of commodity.
Antonovitz & Roe (1984)	Value of rational expectations fore. in fed. beef market	—	Supply response is function of expected mean & variance of price	Fore. of mean and variance of market price	Market	Marshallian surplus	Av. *ex ante* value of rational expectations fore. vs. ARIMA fore. was $0.21 per cwt. over 1970–80 in 1972 $.
Bradford & Kelajian (1978)	Value of wheat crop fore. infor. in US	—	Wheat inventory adjustments	Wheat crop fore.	Market	Marshallian surplus	Point estimate of annual loss to US economy of less than perfect wheat crop fore. is $64 million (1975 $).

Hayami & Peterson (1972)	Marginal social returns to improved crop & livestock statistics	—	Prod. & inventory adjustments	Reduced sampling error contained in USDA survey	Market	Marshallian surplus	Marginal benefit–cost ratios associated with 0.5% reduction in sampling error found to be between 600 to 9 & 100 to 9.
Adams et al. (1995)	Value of improvements in El Niño fore. to US agriculture	Prec. & temp.	Various crop yields	Improvements in accuracy of fore. relative to perfect info.	Market–regional	Marshallian surplus	Free-market value of perfect info. is $144.5 million and of improved fore. $96 million (1990 $).

Clearly, considerable scope exists for improving estimates of the social worth of weather information. In addition to the theoretical problems identified in previous sections, market valuation studies have been hampered by the fact that weather information is a publicly available good, and that diverse groups typically use the same types of weather information. This means that it is frequently difficult to account for all the benefits that result from even a single weather-related prediction.

4. Valuation puzzles

The review of selected empirical value-of-weather information studies and the brief sketch of the theory have suggested several problems or puzzles. If resolved, they represent opportunities for improving the design of weather information systems. The discussion of these puzzles or issues is not intended to imply that the process of providing a more systematic basis for valuing weather information is beyond reach, but instead that there is a broad opportunity for improvement. Of course, in any decision context the objective is to proceed with the best available data and theoretical concepts. They will never be free from uncertainties and assumptions. Changes in the perceptions of the valuation problem, the institutional setting, available technology, and many other critical factors will keep valuation analysis in a constant state of transition.

4.1. Valuation and information processing

From the brief survey of applied valuation studies, it is apparent that weather information is rarely used in the form observed, reported, or even necessarily intended. That is, the weather data as observed and recorded and forecasts as communicated have time and space dimensions. Frequently, transfer functions are applied to these data and forecasts to process them into a form consistent with individual information requirements. Examples include crop yield models, wind chill indices, short-term and long-term forecasts, degree days, soil moisture, and soil temperature. In assigning a value to weather information, it is important to recognize that there can be a confounding of the "information" implicit in the transfer functions with that of the observed or partially processed weather data provided by the system under study. That is, the value of weather information includes the value of the raw observations as well as the value of the information that underlies the various processes used to transform the data into a usable form. In fact, it may be impossible to disentangle the value of the original or unrefined information from the value of the transformation processes themselves.

Who should assume responsibility for the development of these transfer functions or information processing models, and how are the priorities for developing these functions to be determined? In a related vein, are the transfer functions necessary because the system has been designed for purposes other than those of the present users? These and other questions are important for valuing weather information and for developing an appropriate scope for both public and private weather information services.

Presently, it appears, for example, that the U.S. weather information system follows a middle of the road approach in regard to these questions. That is, some processing of information and developmental work on transfer functions aimed at particular clients is accepted as the responsibility of the public agency. Alternatively, there are a number of individuals, organizations, and firms privately processing data provided by public information agencies to make it more useful in specific decision contexts. Public organizations, including state-supported universities, also play a role in processing weather information for specific public uses. Clearly, decisions about public and private responsibility for transfer function development and primary data collection will continue to have an important effect on assessments of the value of information, as well as on the design of the national weather information system.

4.2. Ex ante versus ex post valuations

It is necessary to determine prior to investing in an information system whether the benefit will be greater than the cost. The readily available data on market response to weather information, however, are largely *ex post*. Obtaining appropriate (*ex ante*) valuation calculations will be difficult at best, because the majority of applied valuation studies rely on passively generated data (i.e., data not obtained in an experimentally controlled environment).

Common forms of surplus measures (Hayami and Peterson, 1972; Bradford and Kelajian, 1978) ignore obvious problems associated with using *ex post* secondary data to infer value. It is well understood that actions ranked on the basis of *ex post* measures of consumer and producer surplus cannot be similarly ranked by applying the same measures *ex ante* (Anderson, 1979a; Helms, 1985). These observations on valuation at the market level render many of the existing empirical results on the value of weather information questionable.

Implications of this *ex post* versus *ex ante* conflict are, however, not all negative. Instead, the concepts suggest more constructive ways of proceeding with applied research into the valuation problem. Subjective probabilities regarding plausible states of nature should be elicited from potential users. A new set of subjective probabilities should then be elicited, after the respondents have altered their initial beliefs on the basis of new or additional information. If the message to be received is uncertain, then probabilities associated with possible messages should be estimated as well. In short, problems with *ex ante* valuation simply suggest the use of different types of data and different modeling approaches. These data, in general, cannot be obtained from secondary sources. Investigations using passively generated data are inexpensive but, as is becoming increasingly apparent, provide valuation estimates that are flawed.

4.3. Utility and expected utility maximization

Most applied studies that value weather information in an individual decision context use Bayesian decision-analytic methods (see Section 2.1). Oftentimes such studies have relied upon restrictive utility concepts (e.g., linear utility functions). This is, of course, not a limitation of the Bayesian decision-analytic approach; rather, it reflects the fact that in valuing weather information, utility functions that do not incorporate these restrictive assumptions about attitudes toward uncertainty have only rarely been applied (see Chapter 4 of this volume for some exceptions). There is reason to believe that individuals are generally risk averse (i.e., that utility functions should be concave). Taking into account risk aversion may be an important factor in obtaining more realistic estimates of the value of information. In particular, a risk averse decision maker would not necessarily select the action with the highest expected profit, but rather might prefer another action with lower profit variability. A problem for applied work that attempts to estimate accurately the value of weather information is that if risk aversion is not incorporated, the information may be undervalued (although this is not necessarily the case in general, see Hilton, 1981). Thus, it is necessary to evaluate implications of improved weather information for risk averse decision makers.

4.4. Short- and long-term valuations

Much of the extant theory, as well as the applied work on information valuation, has concentrated on short-term decisions. As is becoming increasingly apparent, though, an important benefit of the weather information system is to provide a capacity for anticipating longer-term events.

The long-term valuation of information is more difficult even within the present theoretical framework. Present versus future trade-offs of benefits and costs provide an especially difficult problem when decisions are made under uncertainty (Chavas, Kristjanson, and Matlon, 1991). For instance, an important problem is to determine appropriate social discount factors. Whose discount factors should be used — those of the present generation or those of future generations? Under what conditions is it possible to aggregate over individual decision-making units to obtain a social discount factor? What is the exact nature of the sequential decision-making process, as well as the associated technological and institutional arrangements, that would allow for long-run adjustments? These and other questions related to the methodology for valuing long-term weather information will lead to a new agenda for theoretical and applied research.

4.5. Distributional effects

Decisions about the public production and dissemination of weather information frequently incorporate distributional concerns. For instance, individuals in agricultural markets serviced by weather information systems may benefit relative to other less well-informed participants. On what basis can society determine the relative utility of benefits generated by the supply of weather information among individuals, as well as among groups of individuals? These are very complicated questions that cannot be easily resolved within the context of the existing economic framework. Many of the individual benefits of information may, in fact, cancel from a societal standpoint. Having better-informed groups of market participants, for example, may simply result in a redistribution of income, leaving total social well-being unaffected. The whole area of social welfare and distributional effects continues to be a prickly issue in economic thought. But for policy decisions, particularly in the design of public services, such problems cannot be dismissed without inviting future difficulties.

4.6. Public versus private information systems

The design of many existing weather information systems is the result of a number of deliberate decisions, natural crises, and other more or less random factors. Current and future discussions will continue to be concerned with the appropriate division between private and public responsibility. These questions on public versus private responsibility are usually evaluated by economists on the basis of efficiency. That is, can public agencies provide information more efficiently than individuals or firms in the private sector? When does the market system fail, resulting in a welfare gain from public intervention? Having established the market, which, if any, components of the public sector's responsibility should be transferred to the private sector? Given the changing technology for sensing and recording weather data and for distributing forecasts from the weather system, these issues will continue to evolve.

The difficulty in determining the appropriate division of private and public responsibility is also related to the transfer function question. All individuals or groups of individuals who are users of weather information must, in one fashion or another, specialize the information to their individual decision needs. These individuals may develop their own personalistic transfer functions or invest in transfer and data processing technology that can be shared. To what extent can this specialization be accommodated given the public–private split in responsibility for weather information services?

5. Concluding observations

Our objective has been to review and assess the results concerning the valuation of weather information and the methods currently used for developing these valuation measures. The first conclusion is that, in both instances, the refinements and extensions in economic decision theory under uncertainty present an opportunity for making important advances in the valuation problem. These advances may lead to workable structures and methods for developing measures of value from a market or societal perspective. From an economic standpoint, these valuation measures require assumptions that are only beginning to be fully understood. A flurry of activity in the market valuation area is likely as the uncertainty and rational expectations theories are more fully merged into information valuation studies.

A second conclusion concerns the measure of information and, perhaps more important, the way data and forecasts from the weather information system find their way into use in economic and social activities. Much of the work on the value of weather information is implicitly related to the value of this information as an input into transfer functions — functions or processes that include extensive types of prior and empirical information not related to weather; for example, plant simulation models. Thus, if an improved basis for valuing the system is to be provided, the role of these transfer functions and their implicit information content must be more clearly recognized and incorporated. The explicit introduction of transfer functions into valuation analysis becomes especially difficult when private versus public information generation and delivery systems are contemplated.

The information valuation problem is difficult. However, the costs of not addressing it directly can be high, both for economists and scientists from other fields, who must ultimately advance the theory of information, and for those directly involved in designing and implementing improved weather information systems.

Acknowledgments

An earlier version of this chapter was presented at the Seminar on the Policy Aspects of Climate Forecasting, jointly sponsored by the National Climate Program Office, NOAA; the National Center for Food and Agricultural Policy, Resources for the Future; and the National Academy of Sciences, Washington, DC.

References

Adams, R.M., Bryant, K.S., McCarl, B.A., Legler, D.M., O'Brien, J., Solow, A. & Weiher, R. (1995). Value of improved long-range weather information. *Contemporary Economic Policy*, **XIII**, 10–19.

Anderson, J.E. (1979a). On the measurement of welfare costs under uncertainty. *Southern Economic Journal*, **45**, 1160–1171.

Anderson, L.G. (1979b). The economics of extended-term weather forecasting. *Monthly Weather Review*, **101**, 115–125.

Antonovitz, F. & Roe, T. (1984). The value of a rational expectations forecast in a risky market: a theoretical and empirical approach. *American Journal of Agricultural Economics*, **66**, 717–723.

Arrow, K.J. (1965). *Aspects of the Theory of Risk-Bearing*. Helsinki: Yrjo Jahnssonin Saatio.

Babcock, B.A. (1990). The value of weather information in market equilibrium. *American Journal of Agricultural Economics*, **72**, 63–72.

Baquet, A.E., Halter, A.N. & Conklin, F.S. (1976). The value of frost forecasting: a Bayesian appraisal. *American Journal of Agricultural Economics*, **58**, 511–520.

Baron, D. (1970). Price uncertainty, utility, and industry equilibrium in pure competition. *International Economic Review*, **11**, 463–480.

Blackwell, D. (1953). Equivalent comparisons of experiments. *Annals of Mathematical Statistics*, **24**, 265–272.

Blackwell, D. & Girshick, A. (1954). *Theory of Games and Statistical Decisions*. New York: Wiley.

Bosch, D.J. & Eidman, V.R. (1987). Valuing information when risk preferences are nonneutral: an application to irrigation scheduling. *American Journal of Agricultural Economics*, **69**, 658–666.

Bradford, D.F. & Kelajian, H.H. (1977). The value of information for crop forecasting in a market system: some theoretical issues. *Review of Economic Studies*, **44**, 519–531.

Bradford, D.F. & Kelajian, H.H. (1978). The value of information for crop forecasting with Bayesian speculators: theory and empirical results. *The Bell Journal of Economics*, **9**, 123–144.

Brown, B.G., Katz, R.W. & Murphy, A.H. (1986). On the economic value of seasonal-precipitation forecasts: the fallowing/planting problem. *Bulletin of the American Meteorological Society*, **67**, 833–841.

Burt, O.R., Koo, W.W. & Dudley, N.J. (1980). Optimal stochastic control of U.S. wheat stocks and exports. *American Journal of Agricultural Economics*, **62**, 172–187.

Byerlee, D.R. & Anderson, J.R. (1969). Value of predictors of uncontrolled factors in response functions. *Australian Journal of Agricultural Economics*, **13**, 118–127.

Chavas, J.-P. & Johnson, S.R. (1983). Rational expectations in econometric models. In *New Directions in Econometric Modeling and Forecasting in U.S. Agriculture*, ed. C.R. Rausser, 205–224. New York: Elsevier–North Holland.

Chavas, J.-P., Kristjanson, P.M. & Matlon, P. (1991). On the role of information in decision making. *Journal of Development Economics*, **35**, 261–280.

Choi, E.K. & Johnson, S.R. (1987). Consumer's surplus and price uncertainty. *International Economic Review*, **28**, 407–411.

Choi, E.K. & Johnson, S.R. (1991). Uncertainty, price stabilization, and welfare. *Southern Economic Journal*, **57**, 789–797.

DeGroot, M.H. (1970). *Optimal Statistical Decisions*. New York: McGraw-Hill.

Ewalt, R.E., Wiersma, D. & Miller, W.L. (1973). Operational value of weather information in relation to soil management. *Agronomy Journal*, **65**, 437–439.

Fama, E.F. (1970). Efficient capital markets: a review of theory and empirical work. *Journal of Finance*, **25**, 383–417.

Freebairn, J.W. (1976). The value and distribution of the benefits of commodity price outlook information. *The Economic Record*, **52**, 199–212.

Gould, J.P. (1974). Risk, stochastic preference, and the value of information. *Journal of Economic Theory*, **8**, 64–84.

Grossman, S.J. & Stiglitz, J.E. (1976). Information and competitive price systems. *American Economic Review*, **66**, 246–253.

Hashemi, F. & Decker, W. (1972). Using climatic information and weather forecast for decisions in economizing irrigation water. *Agricultural Meteorology*, **6**, 245–257.

Hausman, J.A. (1981). Exact consumer's surplus and deadweight loss. *American Economic Review*, **71**, 662–676.

Hayami, Y. & Peterson, W. (1972). Social returns to public information services: statistical reporting of U.S. farm commodities. *American Economic Review*, **62**, 119–130.

Hayek, F.A. (1945). The use of knowledge in society. *American Economic Review*, **35**, 519–530.

Helms, L.J. (1985). Errors in numerical assessment of the benefits of price stabilization. *American Journal of Agricultural Economics*, **67**, 93–100.

Hess, J. (1982). Risk and the gain from information. *Journal of Economic Theory*, **27**, 231–238.

Hilton, R.W. (1981). The determinants of information value: synthesizing some general results. *Management Science*, **27**, 57–64.

Hirshleifer, J. (1971). The private and social value of information and the reward to incentive activity. *American Economic Review*, **61**, 561–574.

Hirshleifer, J. (1989). *Time, Uncertainty, and Information*. Oxford: Basil Blackwell.

Hirshleifer, J. & Riley, J.G. (1979). The analysis of uncertainty and information. *Journal of Economic Literature*, **17**, 1375–1421.

Hirshleifer, J. & Riley, J.G. (1992). *The Analysis of Information and Uncertainty.* Cambridge: Cambridge University Press.

Innes, R. (1990). Uncertainty, incomplete markets, and government farm programs. *Southern Economic Journal,* **57**, 47–65.

Katz, R.W., Murphy, A.H. & Winkler, R.L. (1982). Assessing the value of frost forecasts to orchardists: a dynamic decision-analytic approach. *Journal of Applied Meteorology,* **21**, 518–531.

Kunkel, J.G. (1982). Sufficient conditions for public information to have social value in a production and exchange economy. *Journal of Finance,* **37**, 1005–1013.

Lave, L.B. (1963). The value of better weather information to the raisin industry. *Econometrica,* **31**, 151–164.

Leland, H.E. (1972). The theory of the firm facing uncertain demand. *American Economic Review,* **62**, 278–291.

Malinvaud, E. (1971). A planning approach to the public good problem. *Swedish Journal of Economics,* **73**, 96–112.

Marschak, J. (1971). Economics of information systems. In *Frontiers of Quantitative Economics,* ed. M.D. Intriligator, 32–107. Amsterdam: North-Holland.

McCall, J.J. (1965). The economics of information and optimal stopping rules. *Journal of Business,* **38**, 300–317.

McGuckin, J.T., Gollehon, N. & Ghosh, S. (1992). Water conservation in irrigated agriculture: a stochastic production frontier model. *Water Resources Research,* **28**, 305–312.

McQuigg, J.D. & Thompson, R.G. (1966). Economic value of improved methods of translating weather information into operational terms. *Monthly Weather Review,* **94**, 83–87.

Mjelde, J.W. & Cochran, M.J. (1988). Obtaining lower and upper bounds on the value of seasonal climate forecasts as a function of risk preferences. *Western Journal of Agricultural Economics,* **13**, 285–293.

Mjelde, J.W., Sonka, S.T., Dixon, B.L. & Lamb, P.J. (1988). Valuing forecast characteristics in a dynamic agricultural production system. *American Journal of Agricultural Economics,* **70**, 674–684.

Muth, J.F. (1961). Rational expectations and the theory of price movements. *Econometrica,* **29**, 315–335.

Newberry, D.M.G. & Stiglitz, J.E. (1981). *The Theory of Commodity Price Stabilization.* Oxford: Oxford University Press.

Pesaran, M.H. (1987). *The Limits to Rational Expectations.* Oxford: Basil Blackwell.

Pope, R.D., Chavas, J.P. & Just, R.E. (1983). Economic welfare evaluation for producers under uncertainty. *American Journal of Agricultural Economics,* **65**, 98–107.

Pratt, J.W. (1964). Risk aversion in the small and in the large. *Econometrica,* **32**, 122–136.

Raiffa, H. (1968). *Decision Analysis.* Reading, MA.: Addison-Wesley.

Riley, J.G. (1979). Informational equilibrium. *Econometrica,* **47**, 331–359.

Rogerson, W.P. (1980). Aggregate expected consumer surplus as a welfare index with an application to price stabilization. *Econometrica,* **48**, 423–436.

Roll, R. (1984). Orange juice and weather. *American Economic Review*, **74**, 861–880.

Rothschild, M. & Stiglitz, J.E. (1970). Increasing risk. I: a definition. *Journal of Economic Theory*, **2**, 225–243.

Sandmo, A. (1971). On the theory of the competitive firm under price uncertainty. *American Economic Review*, **61**, 65–73.

Sonka, S.T., Changnon, S.A. & Hofing, S. (1988). Assessing climate information use in agribusiness. II: decision experiments to estimate economic value. *Journal of Climate*, **1**, 766–774.

Sonka, S.T., Mjelde, J.W., Lamb, P.J., Hollinger, S.E. & Dixon, B.L. (1987). Valuing climate forecast information. *Journal of Climate and Applied Meteorology*, **26**, 1080–1091.

Stewart, R.T., Katz, R.W. & Murphy, A.H. (1984). Value of weather information: a descriptive study of the fruit-frost problem. *Bulletin of the American Meteorological Society*, **65**, 126–137.

Stigler, G.J. (1961). The economics of information. *Journal of Political Economy*, **69**, 213–225.

Taylor, C.R. & Talpaz, H. (1979). Approximately optimal carryover levels for wheat in the United States. *American Journal of Agricultural Economics*, **61**, 32–40.

Tice, T.F. & Clouser, R.L. (1982). Determination of the value of weather information to individual corn producers. *Journal of Applied Meteorology*, **21**, 447–452.

Turnovsky, S.J., Shalit, H. & Schmitz, A. (1980). Consumer's surplus, price instability, and consumer welfare. *Econometrica*, **48**, 135–152.

von Neumann, J. & Morgenstern, O. (1944). *Theory of Games and Economic Behavior*. Princeton: Princeton University Press.

Waugh, F.V. (1944). Does the consumer benefit from price instability? *Quarterly Journal of Economics*, **58**, 602–614.

Wilks, D.S. & Murphy, A.H. (1985). On the value of seasonal precipitation forecasts in a haying/pasturing problem in western Oregon. *Monthly Weather Review*, **113**, 1738–1745.

Willig, R.D. (1976). Consumer's surplus without apology. *American Economic Review*, **66**, 589–597.

Winkler, R.L. (1972). *An Introduction to Bayesian Inference and Decision*. New York: Holt, Rinehart and Winston.

Wright, B.D. (1979). The effects of ideal production stabilization: a welfare analysis under rational behavior. *Journal of Political Economy*, **87**, 1011–1033.

4

Forecast value: prescriptive decision studies

DANIEL S. WILKS

1. Introduction

Many human activities and enterprises are affected by uncontrollable future weather conditions. The outcomes of numerous decisions, ranging from fairly trivial (such as opening a window or carrying an umbrella) to vitally important (substantially affecting the livelihoods of one or more individuals or firms), are therefore dependent on these same uncontrollable conditions. It is natural that individuals will seek out information about the future to help make better decisions. That there is substantial interest in, and demand for, weather forecasts indicates that people, at least informally, realize that weather forecasts can be valuable in decision making.

The quality of weather forecasts has gradually improved through time as fundamental knowledge and operational experience has accumulated (e.g., Caplan and White, 1989; Carter, Dallavalle, and Glahn, 1989). Yet it is clear from either informal exposure to weather forecasts, by comparing the local forecast to the ensuing weather in that region or from formal study of the performance of weather forecasts (see Chapter 2 in this volume), that a completely accurate specification of future weather is not possible. There are fundamental physical reasons for this situation, which are described in some detail in Chapter 1. Their implication is that even short-range weather forecasts will never achieve perfection, and that monthly or seasonal forecasts will always be subject to considerable uncertainty. Thus while forecasts can and do reduce the uncertainty associated with future weather events, individuals facing decisions whose consequences will be influenced by these events are now and will always be in a position of uncertainty.

A quite powerful treatment of weather-sensitive decisions made under uncertainty can be accomplished through the construction of what are called decision-analytic models (e.g., Clemen, 1996; Keeney, 1982; Winkler and Murphy, 1985; Winkler, Murphy, and

Katz, 1983). In this approach, described more fully in Chapter 3 of this volume, a decision problem is divided into four basic parts: (i) the actions available to the decision maker, (ii) the possible future unknown events that may occur, (iii) the probabilities associated with those future events, and (iv) specific known consequences following from each possible action-event pair. For the class of problems of interest here, the future uncertain events are meteorological, and the available probabilistic information describing their occurrence is derived.

These four elements are combined in a formal mathematical model of the decision problem under study. The consequences of alternative actions are evaluated with respect to the probabilities of the future events, and actions maximizing the expected (i.e., probability-weighted average) value of some measure of desirability of the outcomes, often monetary, are selected as the optimal decisions. Note that for this analysis to be meaningful and for the forecasts to have value, actions must be available that are capable of producing changes in the consequences. Otherwise the forecasts will offer no more than "entertainment value."

The use of decision analysis to model weather-sensitive decision problems has several advantages. One important result from a decision-analytic model is the specification of the optimal actions associated with particular information about the future: those decisions calculated to provide, on average, maximum benefit to the decision maker. Therefore the approach is sometimes called "prescriptive" in that it specifies, or prescribes, best actions in the face of particular circumstances. The results of a well-formulated prescriptive study are concrete, and can be used as either explicit recommendations or general guidance to help improve decisions made by real managers facing uncertain future events.

Intimately associated with the specification of optimal decisions is the notion of the value of information used in the decision-making process. Existence of a "best" decision implies preferences among at least some of the possible consequences: if a decision maker doesn't care what eventually happens, the decision in question is not a meaningful one. As described in Chapter 3 of this volume, when preferences among consequences can be described entirely by monetary outcomes, computing the value of information (VOI) becomes particularly simple. The improved decisions attainable through use of the information are associated with better monetary outcomes on average than decisions made without the information, and VOI can be computed as the difference between the two average outcomes. Information value, so computed, should be evidence of the advantages to be gained by the potential information user by changing from conventional, suboptimal decision rules. Use and value of hypothetically improved forecasts can also be modeled within this framework. Thus providers of forecasts can assess potential benefits that might be derived from new forecast products under consideration. Also possible is rational evaluation of a fair price for a user or group of users to pay for forecasts tailored to their needs.

On the other hand, while prescriptive models specify what optimal actions should be, these prescribed actions may or may not correspond well to the behavior of real decision makers. The complexity of a particular problem may be so great that the simplifications necessary to achieve computational tractability seriously compromise the relevance of the analysis to its real-world counterpart. That is, aspects of the decision-analytic model may be specified incompletely or even incorrectly. Even assuming that a prescriptive model is well constructed, there is no guarantee that real decision makers will behave optimally in practice. For example, they may be unaware of the superiority, or even the existence, of some of the available actions.

Chapter 5 of this volume reviews "descriptive" studies of the value of meteorological forecasts, which are concerned with analysis and description of the ways in which real decision makers actually use information and make decisions. Even though there will not always be agreement between optimal actions derived from prescriptive analyses and actual human behavior as understood through descriptive studies, the two approaches are complementary. For a prescriptive analysis to be meaningful, it should resemble the real-world decision process reasonably closely. Some rudimentary understanding of a problem is essential before it can be represented in a formal decision analysis. Thus a descriptive analysis could play an important role in the process of constructing a prescriptive model. Once the prescriptive decision model has been constructed, its results have meaning only in the context of the real-world problem. Thus some descriptive understanding of a given decision setting is also an essential component of the evaluation of a prescriptive decision analysis. Conversely, comparison of results from prescriptive and descriptive studies can suggest better or fuller use of available information than is conventionally achieved.

This chapter reviews prescriptive studies of "real" decision situations. This is in distinction to Chapter 6 of this volume, which focuses on "idealized" or prototype situations. Here, in addition to the probability structure associated with a weather/climate forecast setting, sets of potential decisions and their outcomes must be incorporated in a realistic and consistent way.

Actions available to particular decision makers will in general depend on the specifics of the problem at hand. In addition, it can be shown that there are no general monotonic relationships between information value and either the flexibility of actions, the valuation structure of the possible consequences, or the decision maker's level of wealth (Hilton, 1981). Therefore the literature necessarily consists of individual case studies rather than more general results.

2. Case study attributes

In Section 3, existing prescriptive case studies involving the use and value of meteorological forecast information are characterized according to a number of attributes and reviewed. Excluded from the review are studies that treat weather- or climate-sensitive decision problems but use only climatological, as distinct from true forecast, information. Many such studies are reviewed by Halter and Dean (1971), Kennedy (1981, 1986), and Omar (1980). Also excluded from consideration are studies using weather forecasts, but imposing prespecified decision rules rather than internally prescribing the optimal actions (e.g., Gupta et al., 1990a, b; Hashemi and Decker, 1969).

Certain classes of decision problems have been favored in the case studies. Analyses treating agricultural decisions are by far the most common. This should not be surprising, given the importance of food supply to human existence, the exposure of crop plants to environmental (prominently atmospheric) conditions, and the need for most crop plants to be managed and protected to some degree in order to produce economically viable yields. Within the class of agricultural decision problems, surprisingly few areas have received most of the attention. These are raisin production, frost protection, forage preservation, irrigation, crop choice, and fertilizer management. Of the existing nonagricultural studies, most pertain to only two general areas: forestry and transportation. The tables presented in Section 3 summarize the published work to date for each of the above areas, with respect to (i) overall structure of the decision problem, (ii) characteristics of the forecast information, and (iii) information valuation. These three attributes are described in the remainder of this section.

2.1. Structure of the decision problem

Two aspects of the structure of each decision analysis are categorized in the tables. The first of these is the nature of the decision to be considered; that is, the actions available to the decision maker. As indicated above, this is one of the four fundamental elements of any prescriptive decision study.

The second aspect of problem structure tabulated pertains to the relationship in time among decisions in a given setting. The analyses are classified as either "static" or "dynamic." Static analyses are by far the simpler of the two, and the majority of studies to date have adopted a static framework. A static framework implies either that the decisions to be made are single, isolated situations; or, if made in a sequence, that choices and outcomes of a given decision do not affect either available options or possible consequences for subsequent decisions. In either case each decision can be analyzed in isolation. As outlined in Chapter 3 of this volume, probability-weighted average outcomes are computed for each available action, and the action leading to the most desirable expected outcome for that single decision is chosen as optimal. Note that while many problems are well represented as static situations, others that have been analyzed as static problems conform to this model only approximately. Examples also exist of problems clearly not appropriate for analysis in the static framework, but which have been treated as such because of the ease and simplicity of this approach.

Some decision problems are inherently sequential in nature. Actions taken and events occurring at a given point in a sequence of decisions affect options, as well as the physical, biological, and economic consequences of those options, at later stages. Such problems are often referred to as "dynamic," in recognition of the fact that the interrelatedness of changes in their characteristics through time is an essential feature of their structure. It should not be surprising that such problems can become quite complex, and the analysis framework necessary to treat them is more elaborate than that appropriate for static problems.

The basic approach to solution of dynamic decision problems can be understood with reference to a graphical representation of the sequential decision process known as a "decision tree." Figure 4.1 shows a decision tree for a two-stage sequential problem. That is, there are two times at which decisions are to be made. Here squares indicate decision nodes, at which one of two decisions is chosen on the basis of probabilities pertaining to the subsequent uncertain events. These probabilities could be provided by forecasts. The circles represent event (or chance) nodes and, in this simple example, are restricted to two possibilities. At the end of the two decision stages, one of the $2^4 = 16$ terminal points on the right-hand side of the figure will be reached, depending on which choices were made at the decision nodes, and which events subsequently occurred at the chance nodes. Of course, a real decision problem may have more than two actions available at decision nodes, and more than two uncertain events at each chance node. It is assumed that the decision maker knows what each of the 16 terminal consequences is, and assigns a value (often monetary) to each. Since it cannot be known in advance which terminal outcome will occur, the problem is to find the action at each decision node leading to the largest expected (probability-weighted average, using probabilities derived from a forecast) outcome.

A seemingly odd feature of sequential decision problems is that they are solved backward. This is sometimes called "backward induction," or "averaging out and folding back" (Winkler, 1972; Winkler and Murphy, 1985). Consider the two actions, A_1 and A_2, leading to the four terminal consequences $C_1, \ldots, C_4$. The expected return following decision A_1 is simply the probability-weighted average of C_1 and C_2; that is, $\mathrm{ER}(A_1) = p_1 C_1 + p_2 C_2$. Similarly, $\mathrm{ER}(A_2) = p_1 C_3 + p_2 C_4$. Clearly, if the decision process has reached the node from which A_1 and A_2 originate, the optimal action is A_1 if $\mathrm{ER}(A_1) > \mathrm{ER}(A_2)$ and A_2 if the reverse is true. When the same computation is performed for each of the other three possible decisions in period 2, an optimal expected return will be associated with all four of these decisions. At that point

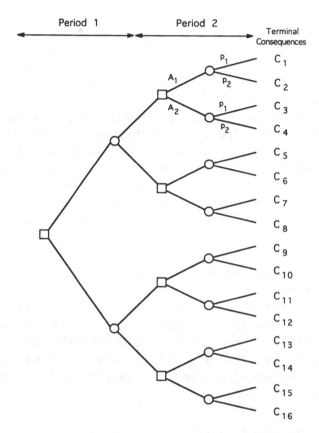

Figure 4.1. A simple decision tree representing a two-stage sequential problem. Squares indicate decision nodes, and circles indicate event nodes. Time runs from left to right; however, the problem is solved in reverse, from right to left.

it is possible to perform the analogous computation for the first decision (leftmost node), using the expected returns computed for each of the 4 second-period decisions where the terminal consequences had been used previously. The approach leads easily to the choice of optimal actions and, for sufficiently simple problems, can yield analytical expressions for the value of information (see Chapter 6 of this volume).

For more complex problems, backward induction is often implemented through a computational approach known as stochastic dynamic programming (e.g., Kennedy, 1986). As before, the goal is to maximize expected return over the full sequence of decisions. However, it is not necessary to draw, or even to imagine, the decision tree. This allows analysis of quite complex and realistic problems.

Regardless of the solution algorithm, the tractability of the approach requires that the status of the decision process at a given time can be specified, at least approximately, by the values of a few "state variables." As the solution procedure works backward in time, it is then not necessary to know the complete sequence of decision and event pairs in the preceding periods (which is good, since these will not have yet been computed), but only their cumulative effects as reflected in the values of the state variables. Thinking in terms of a decision tree, this means that there are as many decision nodes in a given decision period as there are possible combinations of values of the state variables. For example, if a problem is described by two state variables, each of which can take on 1 of 10 values at a given decision stage, then there are 100 decision situations needing to be evaluated at that stage for each possible forecast.

If a given state variable combination can be reached by more than one path through the decision tree, however, the calculations are consolidated as redundant decision tree branches are ignored. The disadvantage is that an analytical solution is not available for subsequent analysis, so that backward induction is often preferred for small, simple problems.

2.2. Forecast characteristics

The forecast information used in the studies reviewed here is characterized according to 5 criteria. The first three are nearly self-explanatory. First, the time scale refers to the length of the period to which the forecast pertains, and must correspond at least approximately to the natural time scale of the decision problem under analysis. Typically the time scale is also of the same order as the lead time, or projection into the future made by the forecast. Second, the predictand is simply the meteorological element or event being forecast. Third, forecast format refers here to whether the forecasts are probabilities (probabilistic forecasts) or statements that one event will occur to the exclusion of others (categorical forecasts).

Clearly the computation of expected returns requires probabilistic forecast information, so that use of categorical forecasts necessarily requires some type of transformation to yield probabilities. Probabilistic forecasts can be used at "face value" in the model computations if they are reliable (that is, well calibrated), in the sense that forecast probabilities correspond well to subsequent event relative frequencies. Otherwise probabilistic forecasts need to be transformed in some way as well. In all these cases, forecast verification (see Chapter 2 of this volume) plays a central role in derivation of the necessary event probabilities, conditional on the forecasts. The most general approach to computing the conditional event probabilities is use of Bayes' theorem, as described in Chapter 3. If it can be assumed that the decision maker's prior beliefs (i.e., information available before receiving the forecasts) are well represented by the climatological probabilities, then direct estimation of the relevant conditional event relative frequencies from a sufficiently large verification data set gives the same result as use of Bayes' theorem. In either case the basic information is contained in the joint distribution of the forecasts and the observations.

Forecast "type" indicates primarily the realism of the forecast information employed in each study. Idealized forecasts are those without a counterpart in current forecasting practice. They may pertain to meteorological events not presently forecast, lead times beyond those in operational use for a given predictand, or both. While studies constructed using idealized forecasts are clearly not useful for improving present-day decision making, they may provide information regarding sensitivity of particular decision problems to different kinds of forecast information, or potentially desirable directions for future forecast product development.

Forecasts denoted as "realistic" are based on actual, operationally available forecast products. Most often in the case studies appearing to date, these are forecasts produced by the U.S. National Weather Service (NWS), although forecasts from other sources are also suitable for use in this type of analysis. In addition to the model forecasts pertaining to the same variable(s) at the same lead time as some real-world forecast product, "realistic" forecasts exhibit statistical characteristics (e.g., accuracy and frequency of use) comparable to their real-world counterparts. For most studies this has meant that a parametric statistical model of the joint distribution of the forecasts and observations has been fitted to a set of forecast-observation pairs. The result is a compact representation of the conditional distribution of the observations given the forecasts, and the predictive distribution (i.e., marginal distribution, or the frequency-of-use) of the forecasts. These are the components of the "calibration-refinement factorization" of the joint distribution of the forecasts and observations (see Chapter 2 in this volume). For example, Katz, Murphy, and Winkler (1982) modeled the joint distribution of forecast and observed minimum temperatures as bivariate normal, with parameters fit using a sample of actual NWS forecasts and subsequent observations. This model yields univariate normal distributions both for the conditional distributions of observed temperatures given each forecast and for the predictive distribution. Note that, defined in this way, use of "realistic" forecasts implies that the climatological distribution of the meteorological event in question is realistically represented. This is because a good representation for the joint distribution of forecasts and observations also implies a good representation for the marginal distribution of the events, through the "likelihood-base rate factorization" (again, see Chapter 2).

A potentially frustrating aspect of decision model construction is that realistic forecasts for a relevant predictand, or forecasts pertaining to a time scale imposed by the decision problem, do not always exist. One solution to this problem is to assume the existence of forecasts as needed, that is, to construct the analysis on the basis of idealized forecasts, as described previously. An alternative that is sometimes possible is to construct transformations of actual forecast products that yield forecasts of the desired form. These are denoted in the tables of Section 3 as "derived" forecasts. For example, many decision problems operate naturally on a daily time scale and may require forecasts of probability distributions for precipitation amounts. Generally available NWS short-term weather forecasts provide probabilities of precipitation occurrence (PoP forecasts) for 12-hour periods. This difficulty was accommodated by Wilks et al. (1993) by combining consecutive pairs of 12-hour PoP forecasts into single 24-hour PoP forecasts (Wilks, 1990a), and then inferring distributions for precipitation amounts on the basis of the PoPs and the climatological precipitation amount distributions (Wilks, 1990b). Once the derived forecasts are constructed, one can tabulate verification distributions for them using available records of forecast performance for the underlying operational forecasts.

It is often of interest to investigate the sensitivity of a given decision problem to changes in forecast quality. That is, in what ways would optimal actions differ if hypothetically improved forecasts were to become available, and what would be the implications of these improvements for forecast value? Studies investigating these quality/value questions are indicated in the "quality changes" rows of the tables in Section 3. This issue is most easily approached when a statistical model of forecast performance is used in the formulation of the decision problem. In that case forecast quality can be varied through adjustments of some of the parameters describing the joint distribution of the forecasts and observations. One approach, for example, is to increase the variance of the predictive distribution to simulate improved forecasts, and to decrease the variance of the predictive distribution to simulate degraded forecasts. In general, the variance of the predictive distribution of climatological forecasts will be zero, since the same forecast is issued on every occasion. At the opposite extreme, the variance of the predictive distribution of hypothetically perfect forecasts will be the same as the variance of the quantity being forecast, since the forecasts and observations match exactly.

It is then possible to present forecast value as a function of hypothetically changed forecast quality. For example, Figure 4.2 shows contours of quality/value surfaces as a function of the standard deviations of the predictive distributions of 90-day temperature and precipitation forecasts, for a choice-of-crop problem at four locations (Wilks and Murphy, 1986). In this case, limiting values of the standard deviation of the predictive distributions, corresponding respectively to climatological forecasts and perfect forecasts, bound the allowable range. Note also that if the predictive distribution is adjusted, corresponding changes in the joint distribution of forecasts and events are necessary for the implied climatological event distribution to remain unchanged.

2.3. Information valuation

One of the primary motivations for employing the decision-analytic framework is the desire to compute potential information value. This question can be thought of as calculating the maximum sum a decision maker should be willing to pay for the forecast information, assuming that the individual would act optimally. As

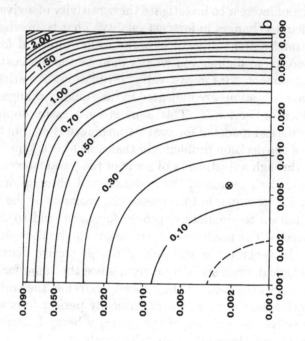

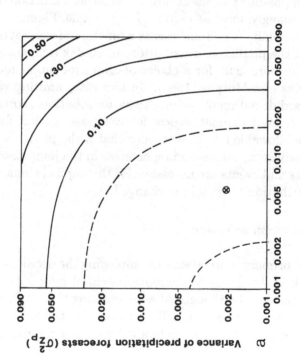

Variance of precipitation forecasts (σ^2_p)

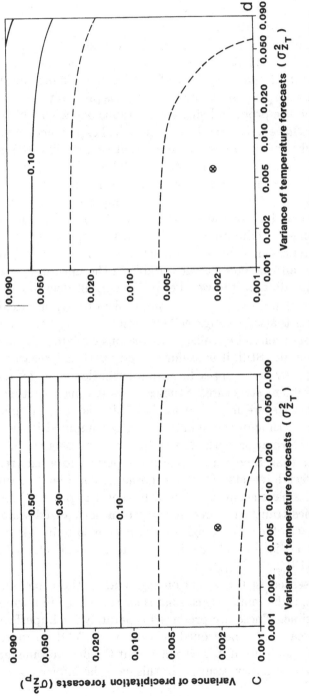

Figure 4.2. Contours of quality/value surfaces for a crop choice problem in four counties in North and South Dakota, using 90-day temperature and precipitation forecasts. The four panels represent a transect from north central North Dakota (a) to southeastern South Dakota (d). Contour interval is $0.10/ha-yr (solid contours), with the $0.05 and $0.01 contours dashed. Circled "x" indicates current forecasts. (From Wilks and Murphy, 1986)

indicated above, this approach implies the existence of a baseline information source available to the decision maker in the absence of the forecast information under study.

Most commonly the information baseline adopted is climatological information. That is, the decision maker is assumed to know the historical relative frequencies for the meteorological events affecting the enterprise under analysis. This is probably a reasonable choice when the problem being considered is one with which decision makers have a reasonably long historical experience. A similar but somewhat more sophisticated alternative is "calibrated persistence," or a "conditional climatology." This information baseline recognizes the tendency for consecutive time periods to exhibit similar weather, and thus may be more appropriate for problems operating on a daily or shorter time scale. Rather than employing a single long-run climatological distribution, "persistence" information is represented as one of a set of conditional distributions for the relevant meteorological variable in the upcoming decision period, depending on the weather in the current decision period.

Often the response of the decision problem to hypothetical perfect forecasts is also investigated. Of course it is impossible for perfect forecasts ever to be realized, for reasons explained in Chapter 1 of this volume. Still, it is useful to study perfect forecasts, since these must provide an upper limit to information value for a given decision problem structure. Studies investigating hypothetically perfect forecasts are indicated in the tables as using a "perfect" baseline, even though the quantity usually computed is the value of perfect information with respect to the climatological information, rather than a negative value for imperfect forecasts relative to the perfect forecasts. One can, however, consider that imperfect forecasts have negative value with respect to perfect forecasts. Some studies transform forecast value to a normalized scale running from zero for climatological information to unity for perfect forecasts; that is, the relative increase in expected return ($\text{ER}_{\text{fcst}} - \text{ER}_{\text{clim}})/(\text{ER}_{\text{perf}} - \text{ER}_{\text{clim}})$.

Also presented in the valuation segments of the tables in Section 3 are specific monetary estimates for the value of information, where possible. These are presented separately for imperfect and perfect forecasts. As mentioned previously, the VOI estimates are very easy to compute if it is assumed that the decision maker acts to maximize monetary returns. In this case the information value is simply the arithmetic difference between the expected monetary return given optimal actions in response to the forecast information and the expected monetary return if only the baseline (e.g., climatological) information is available. Studies adopting this simple maximization of expected monetary value are indicated by "EV" (for expected value) in the "risk treatment" rows of the tables.

Computation of optimal actions through maximization of expected, usually monetary, value amounts to a tacit assumption that the decision maker is "risk neutral." Risk neutrality implies that the decision maker regards the intrinsic worth of money as directly proportional to its amount. For example, risk neutral individuals would be indifferent between the choice of receiving $1 with certainty, or $2 on the flip of a fair coin. Illustrated in this way, risk neutrality may appear to be a quite reasonable assumption and, indeed, many people are essentially risk neutral when small sums are involved. But consider what your reaction would be if offered the choice between $1 million with certainty, or $2 million on the flip of a coin. Unless you are already very rich, the sure million will probably look much more attractive. That is, $2 million is worth less than twice as much as $1 million from your personal perspective, indicating that you are risk averse.

It should be clear from this example that real-world decision makers might very well be sensitive to risk, at least when important decisions are concerned, and that ignoring this sensitivity could produce misleading results in some cases. It is possible to incorporate different risk attitudes into the decision-analytic framework by using the concept of a "utility function" (e.g., Winkler, 1972; Winkler and Murphy, 1985; and Chapter 3 of this volume). A utility function amounts to a mathematical transformation from the original, often monetary, scale to the utility scale, reflecting more fundamentally the perceived worth of each outcome to the decision maker. In terms of the example in the previous paragraph, the utility function, U, of a risk averse individual would satisfy the inequality $U(\$1,000,000) > U(\$2,000,000)/2$. Of course, many functions exist that satisfy this constraint, and this complicates the problem of including risk preferences in realistic decision models. Different individuals will have quite different utility functions (e.g., Baquet, Halter, and Conklin, 1976; Lin, Dean, and Moore, 1974; Hildreth and Knowles, 1986; Winkler, 1972), and the util-

ity function of a single individual might change from time to time
with their circumstance or even their mood. As a consequence,
many studies that maximize expected utility do so using idealized, but (it is hoped) representative, utility functions rather than
those elicited from real decision makers. The mechanics of accounting for risk preferences within the decision model is conceptually straightforward, once the utility function has been defined,
although the estimation of information value is more complex (e.g.,
Hilton, 1981). The computations are performed as described previously, except that decisions are prescribed that maximize expected
utility, rather than expected monetary value. Studies where utility
representation of risk attitude has been employed are indicated in
the tables by "EU" (for expected utility).

3. Case study tabulations

In this section, the existing case studies are tabulated according to
the criteria explained in Section 2. Similar decision problems have
been grouped together in each table. Most of the effort to date has
been directed toward agricultural decision problems, and these are
summarized in Tables 4.1 through 4.7. Forestry-related problems
are tabulated in Table 4.8, and other, primarily transportation
problems, are summarized in Table 4.9.

3.1. Raisins

Two early studies, summarized in Table 4.1, treat the problem of
management of grapes for raisin production in the San Joaquin
Valley of California. The decision pertains to avoiding rain damage during the sun-drying of grapes to preserve them as raisins in
the fall. Kolb and Rapp (1962) consider protecting the raisins from
rain damage during the drying process as a cost–loss ratio problem. This setting, treated more fully in Chapter 6 of this volume,
involves the choice between protection against adverse weather
(rain, in this application) at a cost, versus the loss sustained if
adverse weather occurs without protection. In its simplest, static
form, protection is optimal if the probability of adverse weather
exceeds the ratio of the cost to the loss. Since the drying process
extends over a period of weeks, the daily decision modeled by Kolb
and Rapp should properly be a sequential problem. However, the

Table 4.1. Raisins: case study characteristics

	Kolb & Rapp (1962)	Lave (1963)
Structure:		
Decision	Protect from rain	Harvest timing, raisins or juice
Dynamics	No	Yes
Forecasts:		
Time scale	Daily	3-week
Predictand	Rain occurrence	Rain onset
Format	Categorical	Categorical
Type	Realistic, derived	Idealized
Qual. changes	No	No
Valuation:		
Baselines	Clim. & perf.	Clim.
VOI, imperf.	$50 to $140/ton raisins	Not reported
VOI, perf.	$65 to $155/ton raisins	$225/ha-yr
Risk treatment	EV	EV
Comments:	Problem fit to the cost–loss framework. Categorical forecasts tailored to trigger the optimal decision.	Qualitatively consider market effects; conclude that negative aggregate value to producers is possible.

nature of the fall climate in this region is that precipitation is rare. Accordingly, the static framework is almost appropriate, since the likelihood of overprotection early in the decision sequence is slight.

Lave (1963) considers the somewhat longer-range problem of deciding whether and when to attempt making raisins, versus selling the grapes for juice. Significantly, effects of the use of forecast information on the availability of the local grape processing facilities and on the markets for grape products are considered, although in a qualitative way. This is in contrast to most of the studies reviewed in this chapter, which consider only single decision makers in isolation from the larger economy. Because raisin production is geographically localized, all producers experience essentially the same meteorological conditions in a given year, and have available the same forecast information. Lave speculated that, paradoxically, aggregate value of forecast information to producers could be negative. That is, producers might be better off if none of them

had access to forecasts. Babcock (1990) has verified this specula-
tion for conditions corresponding to Lave's analysis. Note, however, that even in this case the forecast information has positive
value in the usual sense, since a producer choosing to ignore the
information used by the competition would be even worse off. Potential effects on consumer benefits are as yet unanalyzed; as well
as implications for commodities whose production is more widely
distributed, so that different producers see different forecasts and
experience different weather.

3.2. Frost protection

Both studies investigating use of weather forecasts in the frost
protection decision are concerned with protecting blossoming fruit
trees (Table 4.2). The basic problem, familiar to home gardeners,
is whether to take some action to maintain plant canopy temperatures above freezing during nocturnal radiative cooling conditions.
Both studies here model the decision to accomplish this by light-
ing heaters on nights when there is sufficiently high probability of
frost, as indicated by the minimum temperature forecast. Since
fueling and servicing the heaters is expensive, the orchard manager
will want to avoid their unnecessary use.

Baquet et al. (1976) consider protection of pears as a collection
of daily static cost–loss problems, rather than a sequence of related decisions, on 60 spring nights. One strength of this study is
that it uses a variety of utility functions, elicited from real-world
orchardists in the study area, and investigates the effects of their
different risk preferences on the decision problem. These authors
also approach the forecast verification aspect of the problem in a
formal Bayesian way, combining empirical distributions for forecasts conditional on observed temperatures with assumed prior
distributions. Most of the results pertain to use of the climatological distribution as the prior, which is equivalent to direct use
of the conditional distribution of observed temperatures given the
forecasts. However, some computations were also made using a diffuse prior (e.g., Winkler, 1972), yielding a substantial estimate for
the value of climatological information to a hypothetical decision
maker operating without the benefit of such baseline information.

Katz et al. (1982) consider essentially the same problem for ap-
ples and peaches, in addition to pears. In this study the problem is

Table 4.2. Frost protection: case study characteristics

	Baquet et al. (1976)	Katz et al. (1982)
Structure:		
Decision	Orchard heaters	Orchard heaters
Dynamics	No	Yes
Forecasts:		
Time scale	Daily	Daily
Predictand	Min. temp.	Min. temp.
Format	Categorical	Categorical
Type	Realistic	Realistic
Qual. changes	No	Yes
Valuation:		
Baselines	Clim. & perf.	Clim. & perf.
VOI, imperf.	$798/ha-yr	$667 to $1,997/ha-yr
VOI, perf.	$1,270/ha-yr	$1,406 to $3,047/ha-yr
Risk treatment	EU & EV	EV
Comments:	Formulated as a static cost–loss problem. Use a variety of real growers' utility functions.	Bivariate normal model for forecasts and observed temperatures. Dynamic programming solution.

properly treated in a dynamic framework. The solution is obtained through dynamic programming using cumulative bud loss as the single-state variable, and protective action is then prescribed as a function of the forecast, the level of damage so far sustained, and the date.

3.3. Forage preservation

The four studies summarized in Table 4.3 model decisions related to the preservation, usually as hay, of vegetative plant materials (forage) intended for livestock feed at a later date. The study of Byerlee and Anderson (1982) is different from the others, treating the problem of how much hay to store against unknown feed requirements in the following year. More stored feed will be required if the following year is dry, and less will be required if the following year is wet. The decision is thus based on very long-range precipitation forecasts.

Table 4.3. Forage preservation: case study characteristics

	Byerlee & Anderson (1982)	Dyer & Baier (1982)	McQuigg (1965)	Wilks et al. (1993)
Structure:				
Decision	Amount to store	Harvest timing	Harvest timing	Harvest timing
Dynamics	No	No	No	Yes
Forecasts:				
Time scale	Annual	Daily & 4-day	Daily	Daily
Predictand	Prec.	Prec.	"Haying weather"	Prec., temp. & evaporation
Format	Categorical	Categorical	Categorical	Categorical & probabilistic
Type	Realistic	Idealized	Idealized	Realistic, derived
Qual. changes	No	Yes	No	No
Valuation:				
Baselines	Clim. & perf.	Clim. & perf.	None	Clim., persist. & perf.
VOI, imperf.	$59/farm-yr	Not reported	Not reported	$94/ha-yr
VOI, perf.	$520/farm-yr	Not reported	Not reported	$140/ha-yr
Risk treatment	EU & EV	EU	EV	EV
Comments:	Risk treated using both utility maximization and mean-variance analysis.	Emphasize trade-off of forecast accuracy and lead time.	Idealized, focusing on different categorical forecast thresholds for different decision makers.	Treats serial correlation. Monte Carlo simulation to compute variances of returns and harvest and timing distributions.

The remaining three studies cited in Table 4.3 are concerned with the day-to-day harvest decision. Here the physical basis of the problem is the requirement that the forage be dried in the field for one to several days before it is suitable for storage. During this time it is subject to damage, if rain occurs. Rain will also extend the drying period, exposing the forage to the potential for further damage. The problem is complicated by the fact that the quality, and thus value, of the undamaged product decreases as harvests are delayed, so that the farmer cannot wait indefinitely for a very high probability of good drying weather.

The studies of Dyer and Baier (1982) and McQuigg (1965) are idealized formulations, with respect to both the assumed forecast characteristics and the imposition of a static framework on this intrinsically dynamic problem. However, these analyses do point out two important aspects of the use of weather forecasts in decision making. The early McQuigg study demonstrated that weather forecasts could be of value in real agricultural decision problems provided integration of biological, economic, and meteorological factors could be achieved. The Dyer and Baier study illustrates the potential trade-off between forecast accuracy and lead time, that is, that less accurate forecasts may be more valuable if received sufficiently far in advance. The Wilks et al. (1993) study formulates the sequence of daily cutting decisions over a growing season as a large (6 state variables) dynamic programming problem, allowing both the number and timing of cuttings to be prescribed by the model. Temperature forecasts are included to control plant yield and maturity, and as input to the derived evaporation forecasts.

3.4. Irrigation

Three studies, summarized in Table 4.4, have considered use of forecast information to optimize irrigation. Here the problem is to minimize economic crop damage produced by soil dryness, balanced against the costs, in labor and energy required for pumping, in addition to possible costs for the water itself. There is a fairly large literature on dynamic programming solutions to optimal irrigation scheduling (e.g., McGuckin et al., 1987; Rhenals and Bras, 1981) not considering weather forecasts. However, the decision maker would like to meet plant water requirements through rain-

Table 4.4. Irrigation: case study characteristics

	Allen & Lambert (1971a, b)	Rogers & Elliott (1988)	Swaney et al. (1983)
Structure:			
Decision	Irrigation timing	Irrigation timing & amount	Irrigation timing
Dynamics	No	No	No
Forecasts:			
Time scale	Daily	Daily	Daily
Predictand	Prec.	Prec.	Prec.
Format	Probabilistic	Probabilistic	Probabilistic
Type	Realistic, derived	Realistic	Realistic
Qual. changes	No	No	No
Valuation:			
Baselines	Conventional decision rule	Clim., persist., perf., & conventional decision rules	Clim.
VOI, imperf.	$10.53/ha-yr	$3.20 to $19.46/ha	−$7.70/ha-yr
VOI, perf.	Not reported	$3.48 to $18.56/ha	Not reported
Risk treatment	EV	EV	EV
Comments:	Cast as a cost–loss problem; no explicit consideration of rainfall amount.	Cast as a cost–loss problem, perf. VOI sometimes less than fcst. VOI because rainfall amounts not explicit.	Forecasts show negative value resulting from improper problem formulation.

fall, if possible, so that day-to-day precipitation forecasts should be of use in avoiding unnecessary irrigations. Another potentially interesting aspect of this problem, not considered by any of the studies summarized here, is that rainfall following an irrigation can lead to undesirable fertilizer and pesticide leaching.

There are serious shortcomings in all three of the cited studies of this difficult decision problem. In each case a static frame-work has been imposed on the inherently dynamic process. Also, all three include only forecasts for precipitation occurrence rather than probability distributions for precipitation amounts. These and other problems lead to some spurious results, such as the value of imperfect forecasts (or even conventional, nonforecast-based decision rules) being greater than perfect forecasts (Rogers and Elliott, 1988), or negative value for forecast information (Swaney et al., 1983).

It is worth noting that none of the three studies model irrigation decisions in arid regions, and for good reason. While irrigation is widely practiced in such places as California, the use of weather forecasts for irrigation management in these settings would be uninteresting. This is because the climatological probability of rain is so low that there is little scope for improving management practices using forecasts in such climates.

3.5. Crop choice

The issue of crop choice is another important weather-sensitive problem in agriculture. Here the problem is either to choose among crop alternatives or to allocate areas devoted to different crops, on the basis of their anticipated absolute and relative performance in the coming year. These problems thus operate naturally on relatively long time scales, and the near-independence of year-to-year weather generally allows use of the static analysis framework. There are many such problems, and the six analyzed to date and summarized in Table 4.5 by no means exhaust the possibilities.

The first two studies summarized in Table 4.5 focus on the response of the overall United States economy, in terms of producers' and consumers' surplus, to planting decisions. The idealized study of Agnew and Anderson (1977) considers wheat production in relation to hypothetical temperature and precipitation forecasts. Adams et al. (1995) consider choices among four crops in

Table 4.5. Crop choice: case study characteristics

	Agnew & Anderson (1977)	Adams et al. (1995)	Katz et al.* (1987)	Tice & Clouser (1982)	Wilks & Murphy (1985)	Wilks & Murphy (1986)
Structure:						
Decision	Wheat acreage	Allocations among cotton, corn, sorghum & soybeans	Wheat vs. fallow	Corn vs. soybeans	Pasture vs. hay	Corn vs. wheat vs. fallow
Dynamics	No	No	Yes	Yes	No	No
Forecasts:						
Time scale	Seasonal	Annual	Seasonal	Seasonal	Seasonal	Seasonal
Predictand	Temp. & prec.	El Niño	Prec.	"Good," "avg.," & "bad"	Prec.	Temp. & prec.
Format	Categorical	Categorical	Probabilistic	Categorical	Probabilistic	Probabilistic
Type	Idealized	Realistic	Realistic	Idealized	Realistic	Realistic
Qual. changes	Yes	Yes	Yes	No	Yes	Yes
Valuation:						
Baselines	Clim. & perf.	Clim. & perf.	Clim. & perf.	Clim.	Clim. & perf.	Clim. & perf.
VOI, imperf.	$10.8 million/yr	$96 to $130 million/yr	$0 to $10/ha-yr	Not reported	$0.00 to $1.40/ha-yr	$0.00 to $0.14/ha-yr
VOI, perf.	$208 million/yr	$144.5 to $265 million/yr	$116 to $197/ha-yr	$3.65/ha-yr	$11.20 to $17.30/ha-day	$0.20 to $2.40/ha-yr
Risk treatment	EV	EV	EV	EV	EV & EU	EV
Comments:	VOI for entire US, focusing on market effects & links to conventional economic analysis.	VOI for entire US, depending on farm programs. Decisions imposed centrally; not made by individual farmers.	Future-value discounting; VOI depends on location; actions are sensitive to price.	Consider only perfect fcsts., 2-stage dynamics. Also choose fertilization level.	VOI & actions depend on relative prices and DM's utility function.	VOI & actions depend on relative prices and location.

*Results of this study are also reported in Brown et al. (1986).

the southeastern United States in relation to forecasts for El Niño, although the optimal actions are taken to be those that maximize national economic welfare, rather than the expected return to the individual producer who would make the decisions.

Tice and Clouser (1982) consider acreage allocation between corn and soybeans. Fertilization management is also examined, which dictates a sequential model. This is an idealized study treating only perfect forecasts for a generalized predictand.

The three remaining studies use the 90-day outlooks issued monthly by the U.S. Climate Analysis Center (Epstein, 1988). Katz, Brown, and Murphy (1987) analyze the choice of whether to plant wheat, or to fallow (plant no crop) in order to conserve soil moisture for the following year, in the relatively arid climate of the northern high plains. The relevant forecast is for 90-day precipitation, and the year-to-year moisture carryover dictates that the decision be analyzed as a sequential problem. This study also includes use of a future-value discount factor, which reflects the cost of borrowed money or, equivalently, value associated with alternative uses of capital devoted to the farming enterprise. This is an important economic reality for sequences of decisions spanning years. Wilks and Murphy (1986) analyze a similar problem, allowing a second crop in addition to the fallowing choice, but ignoring the sequential nature of moisture storage resulting from fallowing. This study uses the 90-day temperature and precipitation outlooks simultaneously. Wilks and Murphy (1985) is the only study of the five to prescribe decisions maximizing expected utility, although real-world crop choice decisions should depend strongly on the risk attitude of the decision maker (e.g., Lin et al., 1974).

3.6. Fertilization

Amount and timing of fertilization are also important considerations in agricultural production. The three studies summarized in Table 4.6 all treat management of nitrogen, which is generally the most heavily applied plant nutrient, and which can be fairly expensive in times of high energy prices. It is often the case that optimum nitrogen levels depend on future levels of a limiting climatic element (e.g., precipitation in a water-limited climate), which will generally not be known at the time the fertilizer is applied. The two early studies of Byerlee and Anderson (1969) and Doll (1971)

Table 4.6. Fertilization: case study characteristics

	Byerlee & Anderson (1969)	Doll (1971)	Mjelde et al.* (1988)
Structure:			
Decision	Amount, on wheat	Amount, on corn; plant density	Amount & timing, on corn; planting & harvest timing
Dynamics	No	No	Yes
Forecasts:			
Time scale	Seasonal	Seasonal	Seasonal to annual
Predictand	Prec.	"Weather"	Prec. & crop-specific indices
Format	Categorical	Categorical	Probabilistic
Type	Realistic, derived	Idealized	Idealized
Qual. changes	No	Yes	Yes
Valuation:			
Baselines	Clim. & perf.	Clim. & perf.	Clim. & perf.
VOI, imperf.	$0.01 to $0.20/ha-yr	$0.17 to $0.54/ha-yr	$0.10 to $10.40/ha-yr
VOI, perf.	$0.12 to $0.89/ha-yr	$9.93 to $17.40/ha-yr	$21.20 to $46.00/ha-yr
Risk treatment	EV	EV	EV
Comments:	VOI varies as a function of initial soil moisture and fertility.	Focuses on future forecast accuracy necessary for forecast value.	VOI depends on forecast quality, lead time, and prices. Lower quality can lead to higher value if provided further in advance.

* Results of this study are also reported in Easterling and Mjelde (1987), Mjelde and Cochran (1988), and Sonka et al. (1987).

address themselves to the problem of supplying enough nitrogen to exploit the climatic opportunities in the upcoming growing season while trying to avoid unproductive overfertilization. These are cast as once-yearly problems, so that the static framework adopted is appropriate.

The Mjelde et al. (1988) study is much more detailed. In addition to the current-year fertilization problem treated in the other two papers, this study considers fall fertilization of the crop to be planted in the following spring. This decision is weather-sensitive because fall-applied nitrogen is subject to loss by leaching or volatilization if winter and/or spring weather is unfavorable. Also considered here are a variety of other tactical farm management decisions, requiring a sequential framework. An interesting lesson from this work is that forecasts for the most critical part of the growing season with respect to yield may have little or no value if, as is the case here with midsummer weather for corn, no decisions are available allowing use to be made of the information. This problem is an example of a weather-sensitive situation that lacks weather-information sensitivity.

3.7. Other agricultural problems

Two agricultural decision problems that do not fit neatly into one of the above six groups are summarized in Table 4.7. Both are representative of classes of problems that could be fruitfully extended to other crops and locations. Anderson (1973) studies the scheduling of harvest dates for peas, with the decision aided by anticipating their ripening rate using temperature forecasts. An interesting facet of this problem is the potential shortage of harvest equipment during warm years, when the scheduling of different fields must be balanced. Analogous problems occur in other agricultural settings as well, and interesting results will probably be forthcoming when examples of this general class of scheduling problem are analyzed in a dynamic framework.

The Carlson (1970) study is the sole example of another important class of weather-sensitive decision problem, that of scheduling pesticide use. In addition to considerations of expense, minimization of environmental pollution from unnecessary or misplaced pesticide applications is increasingly an important management goal. Carlson considers the level of fungicide necessary to respond to

Table 4.7. Other agricultural problems: case study characteristics

	Anderson (1973)	Carlson (1970)
Structure:		
Decision	Schedule pea harvest	Fungicide type and application schedule
Dynamics	No	No
Forecasts:		
Time scale	Daily & 6–10 day	3 weeks
Predictand	Temp.	Prec.
Format	Categorical	Categorical
Type	Idealized	Realistic, derived
Qual. changes	Yes	No
Valuation:		
Baselines	Clim.	Conventional practice
VOI, imperf.	$0.64/ha-yr	Not reported
VOI, perf.	$4.00/ha-yr	Not reported
Risk treatment	EV	EV & EU
Comments:	A cost–loss problem, considers trading off accuracy vs. lead time.	Focus on best decisions rather than VOI. Mean-variance framework for risk aversion.

future disease pressure brought on by weather conditions. Other meteorological factors can be important in pesticide scheduling and amenable to optimization using forecasts, including the effect of windspeed on unwanted pesticide dispersal, and rainfall leading to washoff of applied materials. Although weather forecasts have been incorporated into other pesticide decision-making problems (e.g., Vincelli and Lorbeer, 1988), the method has involved trial and error rather than comprehensive analysis.

3.8. Forestry

The three forestry-related studies cited in Table 4.8 model quite different decision problems. Anderson (1973) considers the possibility of avoiding the expense of graveling temporary logging roads if perfect precipitation forecasts for the upcoming week were to be available. Brown and Murphy (1988) model the allocation of

Table 4.8. Forestry: case study characteristics

	Anderson (1973)	Brown & Murphy (1988)	Furman (1982)
Structure:			
Decision	Gravel logging roads	Allocate fire-fighting resources	Initiate controlled burn
Dynamics	No	Yes	No
Forecasts:			
Time scale	Weekly	Daily	Daily
Predictand	Prec.	"Fire weather"	Temp., wind
Format	Categorical	Categorical	Categorical
Type	Idealized	Idealized	Realistic
Qual. changes	Yes	Yes	No
Valuation:			
Baselines	Conventional practice	Clim. & perf.	Clim., persist. & perf.
VOI, imperf.	Not reported	$6,084 per pair of fires	NA (utility scale)
VOI, perf.	7.6% of road costs	$16,594 per pair of fires	NA (utility scale)
Risk treatment	EV	EV	EU
Comments:	Value of near-perfect forecasts compared to conventional decision rules.	Idealized problem treating a sequence of two dissimilar decisions.	Utilities assessed by interviews with forest managers.

fire-fighting resources between two fires, including the logistics of mobilization.

Furman (1982) addresses the very different problem of deliberately initiating forest fires as part of the management of the forest stand, where favorable weather consists of conditions unlikely to lead to the fires going out of control. This study is interesting in that there is no monetary underpinning to the preferences among outcomes, but rather utilities are elicited directly from forest managers. These utilities reflect preferences balancing several outcome attributes. In addition to incorporating the managers' risk aversion, they include meeting management objectives, containing costs, and minimizing detrimental environmental impacts.

3.9. Other decision problems

Finally, Table 4.9 presents characteristics of the remaining studies, all but one of which analyze different transportation decisions. The exception is the study by Alexandridis and Krzysztofowicz (1985), which analyzes the decision to generate electrical power, attempting to match demands for heating and cooling as anticipated through temperature forecasts. Results are derived both for the usual categorical temperature forecasts, as well as for probabilistic temperature forecasts (Murphy and Winkler, 1974). Forecast value for nonoptimal decision makers, acting as if the imperfect forecasts were perfect, is also computed.

The Nelson and Winter (1964) transportation study is a simple cost–loss example, where freight can be protected against rain damage if the probability of a nontrivial precipitation amount is sufficiently high. The two early studies of Glahn (1964) and Kernan (1975) are similar, in that the decision in each pertains to what forecast should be issued. In these early studies the problem is so framed because the forecast format was constrained to be categorical. In effect, then, the forecaster is also assuming the role of the decision maker. These two studies are also similar in that they use utility functions not mapped from monetary outcomes, as did the Furman (1982) forestry study.

The Howe and Cochrane (1976) study is unique in simultaneously considering related long-run and short-term decisions. The example pertains to municipalities optimizing investments in snow removal equipment, by using decision-analytic models of the daily

Table 4.9. Other decision problems: case study characteristics

	Alexandridis & Krzysztofowicz (1985)	Glahn (1964)	Howe & Cochrane (1976)	Kernan (1975)	Nelson & Winter (1964)
Structure:					
Decision	Electrical power generation	Ceiling height forecast	Snow removal	Declare air pollution alert	Protect truck freight
Dynamics	No	No	No	No	No
Forecasts:					
Time scale	Daily	5 hours	Daily & "long-run"	Daily	Daily
Predictand	Temp.	Ceiling height	Snow amount	Air pollution	"Heavy" rain
Format	Cat. & prob.	Categorical	Categorical	Categorical	Categorical
Type	Realistic (experimental)	Realistic	Realistic	Realistic	Idealized & derived realistic
Qual. changes	Yes	No	No	No	No
Valuation:					
Baselines	Clim. & perf.	Perf.	Perf.	Clim.	Clim. & perf.
VOI, imperf.	$26,600–$27,200/day	NA (utility scale)	Not reported	NA (utility scale)	$6.40 to $10.40/truck-day
VOI, perf.	$38,900/day	NA (utility scale)	$118,000/yr for one city, relative to fcsts.	NA (utility scale)	$18.40/truck-day
Risk treatment	EV	EU	EV	EU	EV
Comments:	Uses both categorical and probabilistic forecasts. Compares results for nonoptimal decision maker, who believes (imperfect) forecasts are perfect.	Decision is the forecast; choose among possible categorical forecasts to maximize airline utility.	Simultaneous treatment of long-term (equipment acquisition) & daily snow removal decision.	A cost–loss application. Optimize probability thresholds for categorical forecasts.	An early, simple cost–loss application.

snow removal problem given each of the possible levels of equipment purchase. This approach is potentially applicable to many other weather-related facility and investment decisions.

4. Concluding remarks

This chapter has reviewed prescriptive studies relating weather forecasts to real-world decisions, as distinct from the idealized studies described in Chapter 6 of this volume. In this context, decision-analytic models have been constructed and studied primarily for two purposes: (i) to examine quantitatively the potential benefits to be derived from existing expenditures on the meteorological infrastructure, or to examine whether possible future expansion of forecast activities would be justified economically; and (ii) to enable users of weather forecasts to make better decisions, by prescribing optimal actions together with a quantitative analysis of why forecasts lead to better outcomes on average.

More emphasis has been placed on the first of these two purposes to date, although in many ways the framework as applied to "practical" problems is better suited to the second. For reasons that follow from the results summarized by Hilton (1981) on the determinants of information value, it is necessary to analyze individual case studies rather than somehow to compute a value accruing to all users of a particular forecast product. Thus the best that can be expected in terms of breadth are models that reasonably represent a particular class of decision makers, faced with similar decisions under similar climatological conditions. A comprehensive analysis (or even listing) of all users of, for example, daily temperature forecasts does not seem feasible.

Even when a decision problem is well-formulated and realistic, two other problems emerge when representing overall forecast benefits by aggregating VOI over the individuals whose decision problem is described. First, it is not clear that any decision maker actually receives full value, as described in a model, from the forecasts. Real-world decision makers are not constrained to act optimally, which suggests that VOI calculations from prescriptive studies may overestimate forecast value. As discussed in Chapter 5 of this volume, however, these VOI calculations may be underestimates for other nonoptimal decision makers. Choice of the "zero-information" baseline will also influence the relationship between

computed and received information value. Using the conventionally assumed climatological probabilities may underestimate VOI
in cases where these are not accurately perceived by the decision
maker (Mjelde et al., 1988), while in some problems a more sophisticated baseline such as persistence may result in more realistic
forecast valuation (Wilks et al., 1993).

The market-level impact of meteorological forecast information
on overall value has been investigated very little to date, although
the potential exists for surprising effects (Babcock, 1990). Much
more work is needed to investigate the relationships of aggregate
and individual VOI estimates as a function of such influences as
geographic distribution of forecast users and the correlation of
their climates. Expansion of the basis of economic valuation from
producers only to the larger society, including consumers, is needed
as well.

Inevitably, decisions must be analyzed on a case-study basis. Often, designing the analytical structure is surprisingly difficult, especially for more realistic decision problems. One often finds that
the available operational forecasts are not in the proper form, do
not pertain to the most appropriate time scale, or do not predict
the proper variable for the enterprise under consideration. The details of a decision setting are necessarily specific to the particular
problem at hand, and an additional constraint is the rather large
computational resources demanded by a realistic model structure
(e.g., Burt, 1982). For some problems it is difficult to isolate
decisions for which only one time scale is important. In addition,
appropriately structured response models may not be immediately
available, in which case original work that ranges into several disciplines is generally required before the formal modeling and analysis
phase can be initiated.

There does seem to be great potential to improve real-world use
of weather and climate forecast information through continued
work on individual case studies. Many of the agricultural studies reviewed here could be fruitfully extended to analyze management of other commodities, or to prescribe actions at other
locations (i.e., for different climates). Problems in such areas as
livestock management, fisheries, and energy supply also seem ripe
for progress. Realistic, and therefore complex, dynamic programming solutions to a variety of interesting real-world problems are
increasingly within reach as the computational constraints lessen.

Results of sufficiently realistic decision analyses should be useful to real-world decision makers, but as an aid to, rather than a substitute for, their individual judgments.

References

Adams, R.M., Bryant, K.S., McCarl, B.A., Legler, D.M., O'Brien, J., Solow, A. & Weiher, R. (1995). Value of improved long-range weather information. *Contemporary Economic Policy*, **XIII**, 10–19.

Agnew, C.E. & Anderson, R.J. (1977). The economic benefits of improved climate forecasting. Tech. Report (to NOAA), Mathematica, Inc.: Princeton, NJ.

Alexandridis, M.G & Krzysztofowicz, R. (1985). Decision models for categorical and probabilistic forecasts. *Applied Mathematics and Computation*, **17**, 241–266.

Allen, W.H. & Lambert, J.R. (1971a). Application of the principle of calculated risk to scheduling of supplemental irrigation. I: Concepts. *Agricultural Meteorology*, **8**, 193–201.

Allen, W.H. & Lambert, J.R. (1971b). Application of the principle of calculated risk to scheduling of supplemental irrigation. II: Use on flue-cured tobacco. *Agricultural Meteorology*, **8**, 325–340.

Anderson, L.G. (1973). The economics of extended-term weather forecasting. *Monthly Weather Review*, **101**, 115–125.

Babcock, B.A. (1990). The value of weather information in market equilibrium. *American Journal of Agricultural Economics*, **72**, 63–72.

Baquet, A.E., Halter, A.N. & Conklin, F.S. (1976). The value of frost forecasting: a Bayesian appraisal. *American Journal of Agricultural Economics*, **58**, 511–520.

Brown, B.G., Katz, R.W. & Murphy, A.H. (1986). On the economic value of seasonal-precipitation forecasts: the fallowing/planting problem. *Bulletin of the American Meteorological Society*, **67**, 833–841.

Brown, B.G. & Murphy, A.H. (1988). On the economic value of weather forecasts in wildfire suppression mobilization decisions. *Canadian Journal of Forest Research*, **18**, 1641–1649.

Burt, O.R. (1982). Dynamic programming: has its day arrived? *Western Journal of Agricultural Economics*, **7**, 381–393.

Byerlee, D. & Anderson, J.R. (1969). Value of predictors of uncontrolled factors in response functions. *Australian Journal of Agricultural Economics*, **13**, 118–127.

Byerlee, D. & Anderson, J.R. (1982). Risk, utility, and the value of information in farmer decision making. *Review of Marketing and Agricultural Economics*, **50**, 231–246.

Caplan, P.M. & White, G.H. (1989). Performance of the National Meteorological Center's medium-range model. *Weather and Forecasting*, **4**, 391–400.

Carlson, G.A. (1970). A decision-theoretic approach to crop disease prediction and control. *American Journal of Agricultural Economics*, **52**, 216–223.

Carter, G.M., Dallavalle, J.P. & Glahn, H.R. (1989). Statistical forecasts based on the National Meteorological Center's numerical weather prediction system. *Weather and Forecasting*, **4**, 401–412.

Clemen, R.T. (1996). *Making Hard Decisions: An Introduction to Decision Analysis* (second edition). Belmont, CA: Duxbury.

Doll, J.P. (1971). Obtaining preliminary Bayesian estimates of the value of a weather forecast. *American Journal of Agricultural Economics*, **53**, 651–655.

Dyer, J.A. & Baier, W. (1982). The use of weather forecasts to improve hay-making reliability. *Agricultural Meteorology*, **25**, 27–34.

Easterling, W.E. & Mjelde, J.W. (1987). The importance of seasonal climate prediction lead time in agricultural decision making. *Agricultural and Forest Meteorology*, **40**, 37–50.

Epstein, E.S. (1988). Long-range weather prediction: limits of predictability and beyond. *Weather and Forecasting*, **3**, 69–75.

Furman, R.W. (1982). The effectiveness of weather forecasts in decision making: an example. *Journal of Applied Meteorology*, **21**, 532–536.

Glahn, H.R. (1964). The use of decision theory in meteorology. *Monthly Weather Review*, **92**, 383–388.

Gupta, M.L., McMahon, T.A., MacMillan, R.H. & Bennett, D.W. (1990a). Simulation of hay-making systems. 1: Development of the model. *Agricultural Systems*, **34**, 277–299.

Gupta, M.L., McMahon, T.A., MacMillan, R.H. & Bennett, D.W. (1990b). Simulation of hay-making systems. 2: Application of the model. *Agricultural Systems*, **34**, 301–318.

Halter, A.N. & Dean, G.W. (1971). *Decisions Under Uncertainty*. Cincinnati: South-Western.

Hashemi, F. & Decker, W. (1969). Using climatic information and weather forecast for decisions in economizing irrigation water. *Agricultural Meteorology*, **6**, 245–257.

Hildreth, C. & Knowles, G.J. (1986). Farmers' utility functions. In *Bayesian Inference and Decision Techniques*, ed. P. Goel & A. Zellner, 291–317. Amsterdam: Elsevier.

Hilton, R.W. (1981). The determinants of information value: synthesizing some general results. *Management Science*, **27**, 57–64.

Howe, C.W. & Cochrane, H.C. (1976). A decision model for adjusting to natural hazard events with application to urban snow storms. *The Review of Economics and Statistics*, **58**, 50–58.

Katz, R.W., Brown, B.G. & Murphy, A.H. (1987). Decision-analytic assessment of the economic value of weather forecasts: the fallowing/planting problem. *Journal of Forecasting*, **6**, 77–89.

Katz, R.W., Murphy, A.H. & Winkler, R.L. (1982). Assessing the value of frost forecasts to orchardists: a dynamic decision-analytic approach. *Journal of Applied Meteorology*, **21**, 518–531.

Keeney, R.L. (1982). Decision analysis: an overview. *Operations Research*, **30**, 803–838.

Kennedy, J.O.S. (1981). Applications of dynamic programming to agriculture, forestry and fisheries: review and prognosis. *Review of Marketing and Agricultural Economics*, **49**, 141–173.

Kennedy, J.O.S. (1986). *Dynamic Programming, Applications to Agriculture and Natural Resources.* London: Elsevier Applied Science.

Kernan, G.L. (1975). The cost-loss decision model and air pollution forecasting. *Journal of Applied Meteorology*, 14, 8–16.

Kolb, L.L. & Rapp, R.R. (1962). The utility of weather forecasts to the raisin industry. *Journal of Applied Meteorology*, 1, 8–12.

Lave, L.B. (1963). The value of better weather information to the raisin industry. *Econometrica*, 31, 151–164.

Lin, W., Dean, G.W. & Moore, C.V. (1974). An empirical test of utility vs. profit maximization in agricultural production. *American Journal of Agricultural Economics*, 56, 497–508.

McGuckin, J.T., Mapel, C., Lansford, R. & Sammis, T. (1987). Optimal control of irrigation scheduling using a random time frame. *American Journal of Agricultural Economics*, 69, 123–133.

McQuigg, J.D. (1965). Forecasts and decisions. In *Agricultural Meteorology. Meteorological Monographs*, 6, 181–188. Boston: American Meteorological Society.

Mjelde, J.W. & Cochran, M.J. (1988). Obtaining lower and upper bounds on the value of seasonal climate forecasts as a function of risk preferences. *Western Journal of Agricultural Economics*, 13, 285–293.

Mjelde, J.W., Sonka, S.T., Dixon, B.L. & Lamb, P.J. (1988). Valuing forecast characteristics in a dynamic agricultural production system. *American Journal of Agricultural Economics*, 70, 674–684.

Murphy, A.H. & Winkler, R.L. (1974). Credible interval temperature forecasting: some experimental results. *Monthly Weather Review*, 102, 784–794.

Nelson, R.R. & Winter, S.G., Jr. (1964). A case study in the economics of information and coordination: the weather forecasting system. *Quarterly Journal of Economics*, 78, 420–441.

Omar, M.H. (1980). *The Economic Value of Agrometeorological Information and Advice.* Technical Note No. 164, WMO No. 526. Geneva: World Meteorological Organization. 52 pp.

Rhenals, A.E. & Bras, R.L. (1981). The irrigation scheduling problem and evapotranspiration uncertainty. *Water Resources Research*, 17, 1323–1338.

Rogers, D.H. & Elliott, R.L. (1988). Irrigation scheduling using risk analysis and weather forecasts. ASAE Paper No. 88-2043. St. Joseph, MI: American Society of Agricultural Engineers.

Sonka, S.T., Mjelde, J.W., Lamb, P.J., Hollinger, S.E. & Dixon, B.L. (1987). Valuing climate forecast information. *Journal of Climate and Applied Meteorology*, 26, 1080–1091.

Swaney, D.P., Mishoe, J.W., Jones, J.W. & Boggess, W.G. (1983). Using crop models for management: impact of weather characteristics on irrigation decisions in soybeans. *Transactions of the American Society of Agricultural Engineers*, 26, 1808–1814.

Tice, T.F. & Clouser, R.L. (1982). Determination of the value of weather information to individual corn producers. *Journal of Applied Meteorology*, 21, 447–452.

Vincelli, P.C. & Lorbeer, J.W. (1988). Relationship of precipitation probability to infection potential of *Botrytis squamosa* on onion. *Phytopathology*, **78**, 1978–2082.

Wilks, D.S. (1990a). On the combination of forecast probabilities for consecutive precipitation periods. *Weather and Forecasting*, **5**, 640–650.

Wilks, D.S. (1990b). Probabilistic quantitative precipitation forecasts derived from PoPs and conditional precipitation amount climatologies. *Monthly Weather Review*, **118**, 874–882.

Wilks, D.S. & Murphy, A.H. (1985). On the value of seasonal precipitation forecasts in a haying/pasturing problem in western Oregon. *Monthly Weather Review*, **113**, 1738–1745.

Wilks, D.S. & Murphy, A.H. (1986). A decision-analytic study of the joint value of seasonal precipitation and temperature forecasts in a choice-of-crop problem. *Atmosphere-Ocean*, **24**, 353–368.

Wilks, D.S., Pitt, R.E. & Fick, G.W. (1993). Modeling optimal alfalfa harvest scheduling using short-range weather forecasts. *Agricultural Systems*, **42**, 277–305.

Winkler, R.L. (1972). *Introduction to Bayesian Inference and Decision*. New York: Holt, Rinehart and Winston.

Winkler, R.L. & Murphy, A.H. (1985). Decision analysis. In *Probability, Statistics, and Decision Making in the Atmospheric Sciences*, ed. A.H. Murphy & R.W. Katz, 493–524. Boulder, CO: Westview Press.

Winkler, R.L., Murphy, A.H. & Katz, R.W. (1983). The value of climate information: a decision-analytic approach. *Journal of Climatology*, **3**, 187–197.

5

Forecast value:
descriptive decision studies

Thomas R. Stewart

1. Introduction

Studies of the value of forecasts necessarily consider — either explicitly or implicitly — the decisions made by users of the forecasts. Most such studies involve both description (how users actually decide) and prescription (how they should decide). The purpose of this chapter is to present the descriptive approach to studying the value of weather forecasts and to compare it with the prescriptive approach that is treated in several other chapters of this volume (especially Chapter 4).

Both descriptive and prescriptive approaches are based on the belief that the value of weather forecasts is derived primarily from their effects on the decisions of individuals engaged in weather-sensitive activities (McQuigg, 1971), and both approaches require decision-making models to assess those effects. The critical differences between the two approaches are the methods used to develop the decision-making models and the criteria employed for evaluating them. Descriptive models are evaluated according to their ability to reproduce the behavior of decision makers. Prescriptive models are evaluated according to their ability to produce decisions that are optimal according to some normative theory of decision making.

The descriptive and prescriptive approaches are compared in Section 2. In Section 3, a representative sample of descriptive studies is classified and reviewed. Selected results from descriptive research on judgment and decision making that apply to weather-information-sensitive decisions are presented in Section 4. Section 5 includes a brief overview of two broad classes of methods for descriptive decision modeling.

2. Comparison of descriptive and prescriptive studies

In this chapter, the value of forecasts is treated from the perspective of an individual decision maker or a group of decision makers with common interests. Important issues arise when estimating the value of forecasts to an industry composed of individuals who all have the same information (e.g., Lave, 1963) or in determining the value of forecasts at the national level (Johnson, 1990), but these issues are not treated here (also see Chapter 3 of this volume).

2.1. Requirements for a complete descriptive study

Davis and Nnaji (1982) list six types of evaluative information needed to estimate the value of a forecast (p. 463): (i) a payoff function and decision rule based on the information (e.g., a rule for issuing a flood warning based on stream gage readings and a payoff function specifying the costs of all possible combinations of forecasts and events); (ii) a conditional probability distribution of the state of nature given the information; (iii) a probability distribution over the information that might be generated; (iv) the average number of information events per unit time; (v) all the users of the information, their decision rules, and their payoff functions; and (vi) the cost of the information.

These requirements apply to both descriptive and prescriptive studies. The key difference is how the user's decision rules (which constitute a decision model) are obtained. In a prescriptive study, the decision model is based on a normative theory (e.g., Bayesian decision theory) that specifies an "optimal" decision rule (e.g., maximize subjective expected utility). In a descriptive study, the decision model is determined by the user's behavior, and the goal is to obtain an accurate model of the user's actual decision process. As a result, a prescriptive study will yield an estimate of the expected value of the forecast to an idealized user who behaves optimally according to a decision criterion derived from the normative theory; a descriptive study will yield an estimate of the actual value of the forecast to a real user who may or may not use information in an optimal fashion.

Figure 5.1 outlines the steps required for a complete descriptive study that results in an estimate of forecast value. Although a literature search described below found no single study that completed all steps, every descriptive study provided information relevant to at least one step.

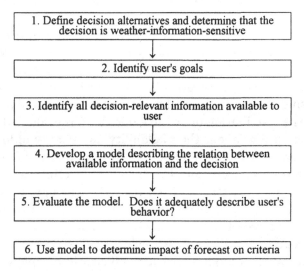

Figure 5.1. The steps in a descriptive study.

The steps in Figure 5.1 are similar to those required for a prescriptive study. In fact, the results of steps 1 through 3 may be similar for both descriptive and prescriptive studies (although in practice this may not be likely). The fundamental differences between descriptive and prescriptive studies are in steps 4 and 5 — model building and evaluation. The standard for evaluating a descriptive model is the user's actual behavior; that is, the model is valid if it adequately describes the user's behavior. Models developed for prescriptive studies need not be evaluated against actual behavior. Discrepancies between prescriptive models and actual behavior are expected because people do not necessarily follow prescriptive models when they make decisions (see Section 4).

Different criteria for evaluating the two types of models dictate different methods for building them. The prescriptive modeler begins with a normative theory, develops the appropriate structure, and collects the data necessary to apply the theory to the decision. The descriptive modeler begins with data obtained from the user's verbal reports and from observations of the user's behavior and, if possible, develops a model that explains the data. Ideally, selection of the type of descriptive model and modeling method is determined by observations and data. In contrast, the prescriptive modeler chooses a model to solve the normative version of the decision problem, not necessarily to match the user's decision process.

To complete step 6, a payoff function must be developed for evaluating decision outcomes, and conditional probability distributions of forecasts and other information must be estimated (or a representative set of scenarios based on historical data could be used). Using the decision model developed in steps 4 and 5, the net payoff of the distribution of possible outcomes is evaluated for different assumptions about the availability and quality of forecasts. Forecast value is computed (assuming the user's utility function is linear in monetary gains and losses) by subtracting the estimate of payoff without the forecast from the estimate of payoff with the forecast. Although this final step is logically required to estimate the value of a forecast, it has rarely been carried out in descriptive studies because, as will be shown below, such studies have rarely resulted in decision models that are sufficiently complete to simulate actual decisions. In contrast, prescriptive studies generally include this step.

2.2. Prescriptive versus descriptive estimates of value

In order to simplify the following discussion, the term "decision" is used to refer to a single action or to a sequential series of actions that may be made in response to a forecast. To clarify the distinction between the expected value of a decision (or series of decisions) and the expected value of information, we will use the term "expected payoff" to refer to the former. The expected value of information is determined by comparing the expected payoff of the decision that would be made with the information and the expected payoff of the decision that would be made without the information. For simplicity here, we assume that utility functions are linear. Methods for estimating value of information when utility functions are nonlinear are described elsewhere in this volume (see Chapters 3 and 4).

If prescriptive and descriptive models are based on the same information and assumptions, the descriptive model will never select a decision that yields a higher expected payoff than the decision selected by the prescriptive model (because, by definition, the prescriptive model chooses the decision with the optimal expected payoff). That does not mean, however, that descriptive

Table 5.1. Economic consequences of decisions in a
hypothetical concrete pouring example

Weather state	Decision	
	Do not pour	Pour
Rain	–$250	–$1,000
No rain	–$250	$1,700

studies will necessarily result in lower estimates of forecast value
than their prescriptive counterparts. Forecast value estimates in
descriptive studies must be based on a comparison of the expected
payoffs of decisions selected with and without the information being evaluated. Because the decisions without the weather information might result in lower expected payoff in a descriptive study
than in a prescriptive study, the difference in expected payoff might
be greater in a descriptive study even though both decisions have
higher payoff in the prescriptive study.

Prescriptive models that are based on an *ex ante* approach to
information valuation cannot result in a negative value of information (before the cost of information is subtracted). In an optimal
model, useless information is simply ignored (i.e., does not change
any decisions) and has zero value. If a descriptive model is used
to estimate forecast value, however, negative values are possible
(Davis and Nnaji, 1982). If, for example, the user misinterprets
a probability forecast, that forecast might turn out to have negative value. The following hypothetical example illustrates how
a descriptive study differs from a prescriptive study and how it
might result in either a negative value of information or a value of
information greater than that obtained from a prescriptive study.

Suppose that Jill is a contractor who needs to pour concrete.
She does not want to pour if it will rain before the concrete sets.
The economic consequences of the possible decision outcomes (in
dollars) are given in Table 5.1.

The prior (or climatological) probability of rain is 0.26, so without a forecast Jill's best choice (assuming she is risk neutral) is to
pour (expected payoff = $998). Suppose that a perfectly reliable
probability of precipitation forecast is available to Jill. Its perfor-

Table 5.2. Statistical properties of probability forecast

Forecast probability	Probability of forecast	Probability of rain given forecast
0.05	0.40	0.05
0.20	0.30	0.20
0.50	0.20	0.50
0.70	0.05	0.70
0.90	0.05	0.90

mance is described in Table 5.2. With the forecast, evaluation of her decision options shows that Jill's best strategy is to pour unless the forecast is for 90% probability of rain. Her expected payoff will then be $1,022, yielding a forecast value (based on prescriptive decision modeling) of $24 per day.

Suppose further that a descriptive model of Jill's decision-making process was developed by interviewing her. In the interview, it was learned that if Jill did not have the precipitation forecast, she would look at the sky in the morning and she would not pour if it were cloudy. This provides a simple descriptive decision rule for the "without forecast" condition:

  (i) If cloudy, do not pour;
  (ii) If not cloudy, pour.

The expected payoff obtained using this rule depends on the conditional probability of rain when it is cloudy (Table 5.3). Since morning cloudiness is a poor predictor of rain, Jill's decision model yields an expected payoff of $596, which is substantially lower than the expected payoff of $998 that she would obtain by not looking at the sky and always pouring.

In the interview, Jill also said that, with the forecast, she would not pour any time the forecast indicated a 50% probability of rain or greater. This rule yields an expected payoff of $899, which is $303 higher than her expected payoff based on morning cloudiness. Therefore, the expected value of the forecast based on the descriptive analysis is $303, which is much greater than the $24 estimate obtained using the prescriptive model.

Another contractor, Jack, said that he would not pour anytime the forecast indicated a 20% chance of rain or greater. Assuming

Table 5.3. *Statistical properties of cloudiness as predictor of rain*

Morning cloud condition	Probability of cloud condition	Conditional probability of rain given cloud condition
Cloudy	0.40	0.35
Clear	0.60	0.20

that he also adopts Jill's "without forecast" approach, his expected payoff would be only $476, which is $120 lower than the expected payoff based on morning cloudiness. Therefore, for Jack, the forecast has negative value.

Because the descriptive value of forecasts is based on the behavior of individuals, the results vary depending on how the forecasts are used. For example, the value of the forecast to Jill and Jack could be changed, and their profits improved, by convincing them to pour unless there is a forecast of 90% probability of rain. They would then have an expected payoff of $1,022, and the estimated value of the forecast *plus* training in how to use it would be $426 (i.e., $1,022 – $596). Of course, the results of prescriptive studies will also vary with characteristics of the individual such as risk preferences and the information available to the individual in the absence of the forecast (Roebber and Bosart, 1996). Chapter 6 of this volume includes a more detailed prescriptive treatment of decision-making models similar in structure to this example.

Because estimates of forecast value from descriptive studies are sensitive to the user's knowledge and sophistication in using the forecasts, the investigator must decide whether a descriptive study will focus on the value of forecasts given current decision-making practices or on the potential value given improved practices that might be implemented through information or training programs. This is also true, at least to some extent, for prescriptive studies because, for example, the results are sensitive to assumptions made about the information used by the decision maker in the absence of forecasts. The effect of such training programs on actual decision-making practices would have to be determined empirically.

2.3. Differences between descriptive and prescriptive studies

In practice, descriptive and prescriptive modelers may bring quite different perspectives, and usually different types of training and experience, to their work. Those perspectives might guide their research in different directions and produce even greater differences in results than the formal differences between their purposes and methods would suggest.

For example, descriptive and prescriptive studies will require different amounts and types of involvement of the user. It might even be possible to do a prescriptive study with no user involvement by relying on secondary data (Sonka, Changnon, and Hofing, 1988). In a typical prescriptive study, one or more users might be interviewed to determine what information is available and to ascertain decision alternatives, constraints, subjective probabilities and utilities, and other parameters required by the prescriptive model. A descriptive study requires more extensive involvement of a representative cross section of users, focusing on the subjective model by which they process information and make choices.

Differences in the results provided by the two approaches can be illustrated by comparing the descriptive study of the fruit-frost problem by Stewart, Katz, and Murphy (1984, hereafter referred to as SKM), based on interviews with fruit growers, the local National Weather Service (NWS) forecast office, and various experts in crop protection with the prescriptive study of Katz, Murphy, and Winkler (1982, hereafter referred to as KMW), based on a dynamic decision-analytic model. Differences are apparent at every step in Figure 5.1.

First, the fruit growers' alternatives were defined in different ways (step 1). KMW considered two alternatives for each night during the frost season: protect or do not protect. SKM found that the important nightly decision was *when* to initiate protection and, of less importance, when to stop protection. KMW's choice was limited somewhat by modeling restrictions. Since they were developing a model that was dynamic over the frost season, their model would have become unmanageably complex if dynamic processes during each night were also included.

Both models recognized that crop protection and minimization of heating costs were the primary goals (step 2), but SKM identified another goal often mentioned by the growers: psychological comfort. Since they faced substantial financial risk during the frost season, growers wanted to have confidence that they could anticipate frost hazard and would be ready to react. Thus, a frost forecast might provide psychological comfort to fruit growers even though their decision to protect their crops might be based on other information. It would be difficult or impossible to assign a monetary value to such a psychological effect, but that does not make it any less important to the user. In a prescriptive modeling approach, such psychological effects could be reflected in the utility function.

The KMW model omitted an item of information (step 3) that growers use to decide when to protect their crops. Almost all large growers have temperature sensors located in their orchards and frost alarms by their beds that are set to wake them when temperatures approach the critical range. SKM found that the frost alarms initiated a period of vigilance that would result in protective measures (usually wind machines) being turned on if temperatures dropped close to the critical level. Consequently, the evening frost forecast was not the trigger for protective action, as assumed in the KMW model.

KMW were able to develop a dynamic decision-analytic model that produced a quantitative estimate of frost-forecast value (steps 4 through 6). SKM had to be content with a qualitative description of the process and were unable to estimate a monetary value for the frost forecast. Although, in principle, a descriptive model could be used to estimate the monetary value of forecasts, this process has rarely been completed. If the objective of a study is to estimate the value of a forecast in monetary units, a combination of prescriptive and descriptive approaches should prove quite valuable, as SKM observe.

Finally, it is important to note that most prescriptive studies necessarily involve descriptive elements. For example, both KMW and Mjelde, Dixon, and Sonka (1989a) developed dynamic models because they recognized that these models better represent the decision processes under study. In their study of crop management, Mjelde et al. assumed that "the farmers are addressing a stochastic intertemporal optimization problem" (p. 1). They solved the crop management problem using stochastic dynamic programming to maximize expected profits with respect to historical climatological probabilities. They found that the static model produced differences in profits ranging from 4% to 10%, relative to the more realistic (i.e., descriptive) dynamic model.

3. Examples of descriptive studies

The DPA Group, Inc. (1985) searched 10 abstract series and data files for studies of forecast value. Their search was updated for this chapter using the sources they found most useful (*Meteorological and Geoastrophysical Abstracts* and the National Technical Information Service). In addition, a bulletin board message was posted on OMNET (a computer network used by atmospheric scientists) asking for information about descriptive studies, and several people active in the field were contacted directly. An annotated bibliography prepared by Mjelde and Frerich (1987) and a review by Mjelde, Sonka, and Peel (1989b) were also examined. This process uncovered only one descriptive study that completed step 6 and resulted in an estimated economic value of a forecast. The remaining studies addressed one or more steps, but few resulted in a complete model that could be used to explain or predict behavior.

Descriptive studies can be roughly classified into four groups: (i) anecdotal reports and case studies, (ii) user surveys, (iii) interviews and protocol analysis, and (iv) decision experiments. Studies representative of each group are briefly reviewed and their contribution to understanding the economic value of forecasts is discussed.

3.1. Anecdotal reports and case studies

Many anecdotes about the value of weather information have been published. For example:

> We interviewed a grain grower who told us how he was able to save thousands of dollars by checking the weather forecast before beginning his harvest. On one occasion, the farmer was about to swath his grain but decided not to do so based on the forecast of precipitation for the following day. The precipitation occurred as forecast and the poor drying conditions persisted for a full three weeks. This farmer figured his grain as No. 3 grain rather than feed. The economic value to him was almost $30,000. (Aber, 1990, p. 55)

Because they suggest an activity for which forecasts have value, such anecdotes might provide the stimulus for serious descriptive or prescriptive studies, but, by themselves, they are of little use for determining the value of forecasts. For example, the value of $30,000 assigned to a single forecast in the anecdote above must be seriously questioned. How accurate is the grower's estimate of his crop's value? How many times did a forecast result in such gains? How many times did it result in losses? What other factors influenced the grower's decision not to cut?

Case studies, usually conducted after decisions have been made, involve more systematic study of the use of weather or climate information than anecdotal reports. For example, Glantz (1982) studied the case of a seasonal water supply forecast for the Yakima Valley, Washington, that turned out to be inaccurate. Based on a forecast of water availability during the irrigation season of less than half of normal, the Bureau of Reclamation allocated 6% of the normal water allocation to holders of junior water rights and 98% of normal to holders of senior rights. In preparation for the predicted water shortage, some farmers had wells dug, leased water rights from other farmers at high prices, and even physically moved their crops to the Columbia River basin where water was more plentiful. Cloud seeding activities were carried out in the Cascade mountains in an attempt to increase snowpack. Actual water availability turned out to be much greater than forecast, and many farmers filed legal actions. Claims for forecast-related losses totaled approximately $20 million.

This case study showed that the forecast clearly did influence the actions of farmers; that is, they took actions that they would not have taken otherwise. It also showed that significant costs could be attributed to the erroneous forecast, and Glantz argued that studies of the value of forecasts should account not only for the value of a good forecast, but also for the potential costs of an erroneous one. In addition to monetary costs associated with the forecast, there are nonmonetary costs. For example, the Bureau of Reclamation lost much of its credibility with farmers, possibly limiting its ability to discharge its responsibilities effectively.

Glantz also pointed out that it is difficult to "separate the influence of the water supply projections from other competing economic influences that might also have been at play in Yakima in 1977" (p. 11). He cited the example of the cattle industry, which was already in trouble and may have been spurred by the forecast to take actions, such as reducing their herds, that they should have taken anyway. It is also possible that actions taken in response to the forecast had longer-term benefits that may have offset their costs. For example, spraying crops for frost protection is more cost-effective than using wind machines, and some of the wells that were dug in anticipation of the 1977 water shortage were later used to supply water for frost protection.

Some case studies include general descriptions of classes of decisions that use weather information. For example, Brand (1992) describes several U.S. Navy decisions that are assisted by weather and climate information. Ryder (1990) reports several uses of weather information in the United Kingdom, and Hawando (1990) presents Ethiopian case studies. Del Greco (1983) describes case studies involving offshore fishing vessels where substantial costs apparently could have been averted if weather forecasts had been used. Changnon and Vonnhame (1986) discuss the use of a seasonal precipitation forecast in a water management problem.

Although difficulty in determining causation is inherent in the case study approach, and the results cannot necessarily be generalized beyond the particular situation studied, case studies do serve an important function in describing forecast value. In depth case studies dramatically show that forecasts are used in complex social, economic, political, and cultural contexts, and that context is an important factor in determining forecast value. Case studies are an important complement to descriptive or prescriptive modeling studies that sometimes must use simplified representations of decision processes in order to develop tractable models.

3.2. User surveys

One approach to studying the economic value of forecasts is simply to ask (by interview, mail, or telephone survey) a representative sample of users how valuable they are. Some surveys (e.g., Easterling, 1986; McNew et al., 1991; Del Greco, 1983) have included questions about whether users are aware of weather services, whether they use them, whether they value them, and how to make the services more valuable. These are essentially marketing studies and may yield useful information for providers of forecasts, such as suggestions for improving services. Other than providing some information about user awareness of weather information, however, they are of little use for deriving realistic estimates of the value of forecasts because they do not reveal how forecasts are used. They should be considered studies of "perceived usefulness" of forecasts rather than their actual value.

Murphy and Brown (1982) review a number of studies of user requirements for short-range weather forecasts. They conclude that two approaches have been taken to the study of user requirements: passive and active. In passive studies, user requests for forecasts are analyzed. The active approach involves the use of a survey method. Murphy and Brown point out that surveys are inherently flawed, because they are based on users' *perception* of needs rather than on either a descriptive or prescriptive decision model.

Some surveys have asked users to estimate the value of forecasts in monetary terms. It is highly unlikely that a user can provide a valid estimate of the economic value of a forecast (see the literature on contingent valuation studies for a list of reasons why, e.g., Fischhoff and Furby, 1988; Mitchell and Carson, 1989). For example, only 8 of 95 people interviewed in one survey of households (Prototype Regional Observing and Forecasting Service, 1979) could provide estimates of the value of weather forecasts in commuting, recreation, and shopping. In their survey of four agribusiness firms, Hofing, Sonka, and Changnon (1987) found that most respondents could not quantify the economic value of better climate information.

Brown and Murphy (1987) sent questionnaires to natural gas companies. One questionnaire requested estimates of savings to the company and willingness to pay for several different types of climate forecasts. They found that many respondents were unable or unwilling to make these quantitative estimates. The estimates that were obtained were highly variable and differed greatly among companies. Of course, such differences might reflect different operating conditions among companies, but it is likely that they are also influenced by the subjectivity involved in making difficult quantitative judgments of forecast value.

Surveys can, however, be used to examine the determinants of *subjective* forecast value. Ewalt, Wiersma, and Miller (1973) incorporated information from 6 case studies of farming operations in 4 Indiana counties lying in different forecast zones and varying in soil type into an interview schedule that was administered to 145 farm operators. Each operator was asked to rate the value of precipitation and field condition forecasts (using a five-point scale) for different field operations, crops, and seasons. Thus, rather than asking users for a quantitative estimate of the monetary value of forecasts, the investigators asked users to rate the relative value of forecasts under different conditions — a much easier and more meaningful task for the user. The rated value of the forecasts was heavily dependent on soil region, crop, field activity (plowing, planting, seedbed preparation, cultivation), and season. Differences in forecast quality across regions were not related to differences in rated value.

The Ewalt et al. (1973) study, like many descriptive and prescriptive studies, underscores the importance of decision-making context in determining forecast value. Although their study was not designed to obtain monetary values, it would not be surprising if monetary values were sensitive to the same kinds of factors that were associated with the operators' rating of forecast value. Both descriptive and prescriptive modeling approaches admit the possibility that, across decision contexts, forecast quality may not be the most important determinant of forecast value.

Other surveys do not address value directly but seek information about how forecasts are used. For example, Curtis and Sites (1987) surveyed 100 recreational anglers along the Strait of Juan de Fuca during the peak of salmon season in the summer of 1986. They found that one-third of the anglers surveyed did not use the available NWS marine warnings and advisories, and 14% were unaware of them. They also found that nearly 90% had never cancelled a fishing trip as a result of a marine forecast. When asked if they ever modified or adjusted NWS forecasts to account for local conditions, 78% responded that they did so at least half the time. The authors speculate that marine warnings make small-boat anglers more alert.

Although it did not produce an estimate of economic value, this survey did provide important information about the value of marine forecasts. It suggested that the decision to cancel a fishing trip is, for many anglers, not weather-information-sensitive. One reason may be that many have limited days off, so that a cancelled trip is lost because the option of delaying the trip a day or two is not available. This result highlights the importance of making a thorough analysis of the user's alternatives before proceeding with a study of forecast value.

The investigators found that 14% of the anglers were unaware of the forecast. For them it had zero value. Apparently, 19% (33% minus 14%) were aware of the forecast but claimed not to have used it. This is a statement about their subjective model (Figure 5.1, step 4), which implies that the forecast has no weight. The finding that many anglers adjust the forecasts is also relevant to step 4. For these people, a descriptive model would have to include a component that represents this forecast adjustment process.

Several user surveys have addressed a critical question that distinguishes descriptive from prescriptive studies. These surveys (Adams, 1974; Murphy et al., 1980; Murphy and Brown, 1983; and Curtis and Murphy, 1985) focused on how forecasts are interpreted by the public. They found (i) that forecasts mean different things to different people, and (ii) that the interpretation of forecasts by the public does not always match what the issuing forecaster had in mind. Mroz and Raven (1993) reviewed literature on public understanding of weather information and reported that research in this area indicates that "weather information in all forms is poorly understood, often misinterpreted, and, at times, even manipulated by the public" (p. 426).

For example, based on a survey of 458 beach users, Adams (1974) concluded that forecasts were underutilized in planning recreational beach trips; he attributed this, in part, to a tendency of users to underestimate the reliability of weather forecasts, to their belief that the forecasts were influenced by an optimistic bias, and to users' misinterpretation of the forecasts. He argued that forecasters should share responsibility for users' misunderstanding of forecasts and should strive to improve communication with the public. Results like these indicate that descriptive models must account for users' interpretations of forecasts.

Although surveys are generally not appropriate for producing descriptive models of user decision making, they may provide useful information to help guide the modeling process. One inescapable conclusion from surveys is that there are important individual differences among users of weather and climate information. As a result, it will generally not be possible to develop a single descriptive (or prescriptive) model that represents all users even in the case of a specific application. For example, the model for those who are aware of the forecast but don't use it will differ from the model for those who do use the forecast, and that model will differ from the model for those who use the forecast but adjust it first. For this reason, descriptive modeling must be idiographic rather than nomothetic (Hammond, McClelland, and Mumpower, 1980, pp. 117–119); that is, models will have to be developed for individual users rather than for user groups. Any modeling process that involves averaging over groups of users without accounting for individual differences is likely to produce misleading results. This problem also arises in the case of prescriptive models.

3.3. Interviews and protocol analysis

In these studies, descriptions of users' decision-making strategies, or "protocols," are developed from verbal reports obtained from interviews or from analysis of other written materials. The SKM study of fruit growers described above is an example of this approach.

Another example is a study by Glantz (1980), who interviewed 60 experts regarding possible responses to a hypothetical El Niño forecast. This study addressed primarily step 1 in Figure 5.1 (alternatives and weather-information sensitivity). He found that political, economic, and other constraints might limit the flexibility of decision makers in Peru or elsewhere to respond to an El Niño forecast. Although he did not make a quantitative estimate of the value of the forecast, Glantz concluded that, even though an El Niño forecast may be useful in theory, "in practice it appears that its value may be quite limited" (p. 449).

Glantz (1977) analyzed the responses of more than 100 people representing "a wide range of fields" in the Sahel regarding what they would have done if an accurate forecast of rainfall and temperature had been available in 1973, a devastating drought year. He organized the responses into those describing "what should be" and those describing "what is." He tentatively concluded that "given the national structures in the Sahelian states in which a potential technological capability would be used, the value of a long-range forecast, even a perfect one, would be limited" (p. 156). He also argued that the distinction between "what should be" and "what is" is particularly useful for analyzing the implications of a long-range forecast because it helps to avoid the pitfalls of adopting either a utopian standpoint or a reality standpoint, and it can be used to "draw attention to the often implicit assumptions concerning the potential benefits for society of new technology" (p. 157).

Changnon (1992) conducted extensive interviews with 27 agribusiness executives regarding their use of climate predictions. He found that climate predictions are seldom used in making major decisions, which implies that they have little economic value. The study identified major impediments to the use of climate predictions and potential strategies for improving their usefulness. Similar results were obtained from a mail survey of 114 respondents reported by Sonka, Changnon, and Hofing (1992) and from interviews of 56 decision makers in 6 power utilities regarding their use of climate forecasts (Changnon, Changnon, and Changnon, 1995).

Occasionally an agency, business, or individual develops and documents an explicit model for the use of weather information. For example, the Federal Aviation Administration (FAA) regulations for the use of terminal forecasts (Mathews, 1992) require commercial air carriers to carry extra fuel if the terminal forecast indicates any possibility of ceiling height below 2,000 feet or visibility less than three miles within one hour of arrival. The FAA regulations can be translated directly into a simple model for the use of the terminal forecast (Figure 5.2). This is an example of one form that a descriptive decision model can take. If the necessary data about conditional distributions of weather events and costs were obtained, the model could be used to estimate the value of terminal forecasts. Of course, estimating costs (e.g., passenger inconvenience) would be no simple matter.

Studies by Suchman, Auvine, and Hinton (1979, 1981) involved an attempt to develop simple decision models and use them with historical data to estimate the economic benefits of forecasts. They contacted business and government clients of a private weather service using a questionnaire and, in some cases, a followup phone call and personal visits. The responses were used to develop simple decision models indicating what the users would do given various forecast scenarios. Using the data on forecasts and actual events for a year, Suchman et al. were able to estimate (retrospectively) yearly "direct economic losses" resulting from incorrect forecasts for various user groups. The monetary values computed were not measures of forecast value, but estimates of yearly loss resulting from imperfect forecasts. The authors pointed out that these values did not measure the value of the weather service relative to

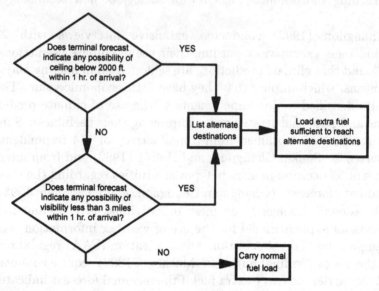

Figure 5.2. Descriptive model of use of terminal forecasts based on FAA regulations.

some realistic alternative; at best, they measured the maximum additional value that could be obtained by improving forecasts. Suchman et al. (1981) attempted to estimate the economic benefit of satellite data by comparing costs of imperfect forecasts during the two years before satellite data were available to the consulting firm with the year after the data became available. The method failed, however, because of large differences in weather during the two periods. The authors note that a much larger sample size would be needed to make a realistic comparison of this type.

As descriptive studies of the value of weather information, the Suchman et al. studies have two major limitations. First, monetary values were estimated for perfect information relative to the actual forecasts, rather than for the actual forecast relative to some alternative such as climatology or publicly available forecasts. Such a procedure is useful for estimating the maximum value of forecast improvements, but not for estimating the value of the forecast itself. Second, the users' decision models were not validated and were almost certainly oversimplified because they did not account for the context of decisions. As the authors point out, the use of their method assumes that users actually do what they report on the questionnaire. The validity of that assumption was not examined. Furthermore, the decision models were based only on the forecasts. That is, the investigators assumed that the forecast alone determined user action, regardless of other information or circumstances existing at the time. This assumption is unlikely to be true for all users.

3.4. Decision experiments

Sonka et al. (1988) have proposed the use of "decision experiments" to assess the value of climate information and provide a study illustrating this approach. They interviewed two key managers responsible for production planning in a major seed corn producing firm. The participants were given 8 climate prediction scenarios (representative of actual growing season climate conditions) and asked what actions they would have taken if they had received this information when planning decisions had been made for the current season. The resulting decision model (Table 5.4) provides another example of the form that a descriptive decision model can take.

This decision model is hypothetical because it represents the managers' responses to perfect climate forecasts that are not currently available (Sonka et al. modeled responses to imperfect forecasts as well). Hypothetical models pose difficulties because there are no objective criteria for validating them; that is, they cannot be compared with actual behavior. It is often necessary to use them in descriptive work, however, because real decisions may not be available. If observations of real decisions are not available as a basis for modeling, then it is important to obtain the data for modeling under conditions that are similar to those present when real decisions are made. In other words, the data for descriptive decision modeling must be obtained under representative conditions. The necessity for representative design has been stressed by Brunswik (1956) and Hammond et al. (1975). Extensive research on descriptive modeling underscores the need for representative design and indicates that properly designed experiments can result in models that predict actual decisions (see Brehmer and Brehmer, 1988; Stewart, 1988).

As in the case of all decision models, the Sonka et al. model operates only within a limited range of circumstances. In this case,

Table 5.4. Decision model for seed corn producing firm

Type of seasonal climate prediction	Implications for corn production	Decision
A	Very serious	Shift varieties toward irrigated areas
B	Normal to favorable	No action
C	Moderately serious	Move varieties from site in east central Corn Belt toward western and northern plants sites
D	Very serious	Move midseason varieties away from areas in central Corn Belt areas
E	Somewhat serious	Move late-season varieties toward irrigated and eastern sites
F	Somewhat serious	Move late-season varieties toward irrigated areas and other eastern sites
G	Moderately serious	Move early varieties from eastern Corn Belt to west central Corn Belt and to three irrigated areas
H	Normal to favorable	No action

Source: Adapted from Sonka et al. (1988), table 1.

the model applies only to 1987 because, except for the climate prediction, all conditions in the experiment were defined by the 1987 crop season. The model also applies only to the range of weather types included in the set of predictions.

In order to estimate forecast value, Sonka et al. developed an empirical model of the firm's seed corn production process. By combining the decision model with the production model, they were able to simulate with- and without-information scenarios for 100 random 10-year scenarios. Information value was then calculated, and sensitivity of the estimated value to model parameters was investigated. This study demonstrates the feasibility of using a descriptive decision model to obtain quantitative estimates of forecast value. It is the only such example found in the literature search.

In another decision experiment, Baker (1984) asked citizens to respond to hypothetical scenarios describing hurricane threats.

For one group of 100 citizens, the scenarios varied on 4 factors: (i) severity of storm, (ii) storm track and position, (iii) National Hurricane Center "alert" status, and (iv) evacuation notice from local officials. A second group received the same scenarios plus a forecast specifying the probability that the storm would cause hurricane conditions. Baker developed logistic regression models to predict the percentage who would evacuate under each scenario. He found that provision of hurricane probabilities (in addition to National Hurricane Center alert information and all the other information) had a negligible effect on the percentage evacuating (around 5%). The most important factor was local officials' advice or orders to evacuate. No payoff function was developed, and forecast value was not estimated.

The descriptive model developed in Baker's study was intended to predict group behavior (i.e., evacuation percentage). As noted above, decision models must account for individual differences among users. Because such differences were not analyzed, the validity of the aggregated model is suspect.

3.5. Summary

Most descriptive studies can be classified as case studies or surveys. Although such studies provide information relevant to forecast value, they do not produce quantitative estimates of value. Studies involving the detailed interviews, protocol analysis, and decision experiments that are required to develop descriptive models for estimating forecast value are rare.

Although descriptive studies have rarely produced estimates of forecast value, they do suggest that users often do not obtain the maximum possible value of forecasts and that the usefulness of forecasts may be limited by: (i) constraints that deny the flexibility to respond to the information, (ii) lack of awareness of the information or ability to obtain it, (iii) misunderstanding or misinterpretation of forecasts, (iv) availability and use of locally specific information more appropriate to a specific decision than the forecast, (v) nonuse or nonoptimal use of forecasts, or (vi) low importance given to weather information because other nonweather factors are judged more important.

In addition, the studies indicate that there are important individual differences in the use of forecasts. Those individual differences may well translate into differences in the value of forecasts to different users, even if they face the same decision with the same payoff structure.

4. Factors that affect differences between descriptive and prescriptive models

Decisions that are likely to benefit from weather and climate forecasts always involve uncertainty. Also, they are usually complex decisions because they involve many interdependent elements, or because several different types of information must be considered before a decision is made, or both. Often, multiple criteria must be satisfied, and trade-offs between competing or conflicting criteria may be necessary. The decisions are often part of a dynamic process; that is, a sequence of decisions is made in the context of a changing environment where each decision depends, in part, on the outcomes of earlier decisions. Thus, weather and climate forecasts are likely to be used in decisions that involve uncertainty, complexity, multiple criteria, trade-offs, and dynamic processes. These characteristics are present in most of the kinds of decisions that interest judgment and decision researchers. For reviews of this research, see Janis and Mann (1977), Kahneman, Slovic, and Tversky (1982), Arkes and Hammond (1986), Hogarth (1987), Baron (1988), Dawes (1988), Brehmer and Joyce (1988), and Payne, Bettman, and Johnson (1992). In this section, a few illustrative results of descriptive research on judgment and decision making that may apply to weather-information-sensitive decisions are described.

Decision making under uncertainty has been a major topic of study in judgment and decision research. The major conclusions of this research are: (i) that people do not make decisions according to the subjective expected utility model (see Chapter 3 of this volume); and (ii) that people often ignore relevant information, are unduly influenced by irrelevant information and, consequently, often perform poorly in such situations. A series of descriptive studies of judgment, inspired by the work of Tversky and Kahneman (1974), has identified a number of mental strategies (heuristics) that can lead to errors in judgments (biases). Research results repeatedly recount numerous biases in judgment, the inconsistent and poorly controlled nature of the judgment process, the pervasiveness of cognitive limitations that can reduce the validity of judgments, and the difficulty of overcoming those limitations.

Although this research clearly demonstrates that prescriptive models do not necessarily describe human behavior, its applicability to actual decisions made by users of weather forecasts is difficult to determine. The generality of the heuristics and biases research conducted by Tversky and Kahneman and many others has been seriously challenged by a number of authors (e.g., Ebbesen and Konecni, 1980; Nisbett et al., 1983; Kruglanski, Friedland, and Farkash, 1984; Gigerenzer and Murray, 1987; Lopes, 1991; Fraser, Smith, and Smith, 1992). In this author's opinion, extreme caution should be exercised in making generalizations from this research to the behavior of decision makers in field (i.e., "real-world") settings. However, the possibility that human cognitive processes for coping with uncertainty are, under some circumstances, flawed cannot be discounted.

Complexity of the decision problem has also been studied in judgment and decision research. Complexity varies with amount of information, interdependence among variables, number of criteria to be satisfied and trade-offs among them, and whether the process is dynamic or static. Not surprisingly, because people are limited in their capability to process information, performance decreases with complexity. One might expect, therefore, that differences between descriptive and prescriptive models will increase as complexity increases.

Another set of studies has focused on the intuitive versus analytic nature of the judgment process. Prescriptive models are highly analytic; that is, they imply explicit problem structuring, formal reasoning, and solution by calculation. Human judgment processes can vary along a continuum from intuitive to analytic (Hammond, 1990; Hammond et al., 1987). Decision makers who use highly analytic processes generally can describe those processes in more detail and more accurately than those whose decision making is more intuitive. Therefore, we would expect greater differences between prescriptive and descriptive models for intuitive decision makers than for more analytic ones.

Although prescriptive studies generally focus on a single criterion for decision success (economic losses or gains), multiple (including nonmonetary) criteria can be handled within a prescriptive model (Keeney and Raiffa, 1976). In descriptive modeling, it is not unusual to find that a decision maker has more than one goal and that decision success is not always easily measured in monetary units. Thus, both prescriptive and descriptive studies may involve multiple criteria that include nonmonetary considerations. When decisions involve multiple criteria that cannot all be satisfied at once, trade-offs are necessary. Research suggests that people avoid trade-offs and prefer to make decisions based on a single dimension (Baron, 1988). In order to do this, they may try to think of reasons to ignore or discount other dimensions (Montgomery, 1984). This treatment of multiple criteria differs from the usual treatment in prescriptive models, which involves some weighted combination of criteria. When multiple criteria are involved, descriptive models will have to account for the decision maker's method of combining or selecting criteria in different contexts.

Studies of decision making in dynamic situations are a relatively recent development in descriptive decision-making research. Hogarth (1981) argued that many of the results obtained in the study of decision making under static situations might not apply to dynamic tasks, and that the heuristics that lead to poor performance in static tasks might produce good performance in dynamic tasks. Several recent studies have focused on delay of feedback, and found that people have difficulty performing well in tasks that involve a time interval between the time an event occurs and the time that they receive the information that it has occurred (e.g., Sterman, 1989; Brehmer, 1990). Further research is needed in order to understand the conditions under which human performance in dynamic tasks can approximate those of an optimal model.

In summary, the descriptive research on judgment and decision making has found important differences between prescriptive and descriptive models in decisions that involve uncertainty, complexity, and multiple criteria. In addition, differences between prescriptive and descriptive models are likely to be greater when the decision process is performed intuitively. The study of dynamic decision making is relatively new, and little is known about the relation between prescriptive and descriptive models for this type of problem.

5. Overview of descriptive modeling methods

Approaches to descriptive modeling of decision processes (step 4 in Figure 5.1) can be classified into two broad groups: (i) protocol analysis (or process tracing), and (ii) judgment analysis. These approaches are briefly summarized here.

Protocol analysis relies on users' verbal descriptions of their reasoning. Protocols are best obtained by asking individuals to "think aloud" while they work through a series of decisions. The protocols are analyzed to develop a model that consists of a number of "if . . . then . . ." relations that can be diagramed in the form of a flowchart. Protocol analysis has been used extensively by "knowledge engineers" to develop computer-based expert systems, and it has a long history in psychology as well (Kleinmuntz, 1968; Ericsson and Simon, 1984). Recently, the use of verbal reports to derive an "influence diagram," which can be considered a form of protocol, has become popular in decision analysis (Oliver and Smith, 1990).

Traditionally, many psychologists have been suspicious of verbal reports, and they have used methods for analyzing judgment that do not depend on a person's ability to provide accurate verbal descriptions of decision processes. Judgment analysis refers to a class of methods for deriving models of decision making by analyzing a sample of actual judgments or decisions. The user is required to do only what he or she does naturally; that is, to make decisions using familiar materials. The analyst develops a model to describe the inference process that produced the decisions.

The data required for the development of such a model are a number of cases of a particular type of decision. Each case includes the information used to make the decision and the resulting decision itself. Cases may be obtained in a natural setting (e.g., a fruit grower's decision to protect crops each night during the frost season) or in controlled settings. In the latter situation, the forecast user would be asked to make judgments based on a sample of real situations or hypothetical scenarios designed to represent real situations.

The items of information available to the user, called "cues," are considered independent variables in an analysis, and the decision is the dependent variable. Multiple regression analysis is one statistical technique that has been used with considerable success to fit a model to the data (Stewart, 1988), but a number of other methods have been applied as well, including analysis of variance and conjoint measurement. The analysis yields a model of the user that expresses the decision as a mathematical function of the cues and an index of how well the model fits the judgments (e.g., the square of the multiple correlation coefficient).

Although the use of statistical models of judgment can be found as early as 1923 (Wallace, 1923), extensive use of the method began in the mid 1950s (Hammond, 1955; Hoffman, 1960). Decision processes of a variety of experts have been modeled, including physicians, stockbrokers, clinical psychologists, polygraph interpreters, and weather forecasters (Stewart et al., 1989; Lusk et al., 1990). Descriptions of a theoretical foundation for judgment analysis and descriptions of the method itself can be found in Hammond et al. (1975) and Brehmer and Joyce (1988).

Judgment analysis offers one major advantage as a tool for developing descriptive models of the use of weather forecasts: It provides a modeling method that does not rely on the user's ability to describe his or her thinking process. This is important because the ability to make decisions is not always accompanied by the ability to describe accurately the process that produced the decisions, particularly when the process contains intuitive elements. Verbal descriptions of reasoning can be incomplete, inaccurate, or misleading. Some important aspects of the decision process may not be readily accessible and may be difficult to translate into words. The description obtained may be unduly influenced by the method used to elicit it. Questions posed by the investigator impose a "frame" on the problem. Seemingly irrelevant and inconsequential aspects of problem framing can have a powerful effect on judgment (Tversky and Kahneman, 1981). For these reasons, it is desirable to have a modeling tool that does not depend on the expert's ability to describe the inference process.

Judgment analysis can be used in combination with verbal protocols (Einhorn, Kleinmuntz, and Kleinmuntz, 1979), so that the investigator can take advantage of the insights provided by both while avoiding their pitfalls. If the investigator relies solely on the user's verbal statements, then both the efficiency of model development and the ultimate accuracy of the model will be limited by the user's ability to describe verbally his or her decision process. If, on the other hand, the investigator relies solely on judgment analysis, certain important, but rare, situations that are not represented in the sample of cases may be misunderstood. Although the two techniques complement each other well, few studies have combined them.

6. Conclusion

In order to understand fully the value of weather forecasts, both prescriptive and descriptive studies are necessary. Prescriptive studies provide estimates of the potential value of forecasts under the assumption that the decision maker follows an optimal strategy. They are relevant to the study of forecast value, but do not necessarily provide reliable estimates of the actual value to society given current decision practices. Descriptive studies can estimate the value of information given current users' decision-making practices, but this approach may overlook the possibility of additional value resulting from improvement in those practices.

Because the approaches are complementary, comparison of the results of descriptive and prescriptive studies of the economic value of the same forecast can be especially enlightening. Assuming that the decision maker is risk neutral (see Chapters 3 and 4 of this volume for a description of methods for calculating the value of information when the decision maker is not risk neutral) and that the expected value of information is calculated by comparing the expected payoff of decisions made with and without the forecast, Table 5.5 describes five possible outcomes of a comparison of descriptive and prescriptive studies. The implications of each of these outcomes are described below.

(i) *Both prescriptive and descriptive studies produce the same expected payoff of the decision made with the forecast.* This would imply that the decision payoff is not sensitive to the decision modeling approach, and an increase in payoff through training users to make better use of information (making the descriptive model more prescriptive) is unlikely. This result would indicate that the prescriptive model was actually a good descriptive model as well, as was found in a study of Sri Lankan rice farmers by Herath, Hardaker, and Anderson (1982). In this case, any differences in the value of information could be caused only by differences in the expected payoff of the "without-forecast" decision. If this expected payoff is lower for the prescriptive study than for the descriptive

Table 5.5. Possible results of comparison of prescriptive and descriptive studies

		Expected value of forecast		
		DEV > PEV	DEV = PEV	DEV < PEV
Expected payoff of decision with forecast	DEP = PEP	(1a)	(1b)	Not possible
	DEP < PEP	(2a)	(2b)	(2c)

Expected value of forecast = (Expected payoff with forecast) − (Expected payoff without forecast)

DEP represents the expected payoff of the "with forecast" decision estimated using descriptive modeling approach.

PEP represents the expected payoff of the "with forecast" decision estimated using prescriptive modeling approach.

DEV represents the forecast value estimated using descriptive modeling approach.

PEV represents the forecast value estimated using prescriptive modeling approach.

study, then the value of information is greater in the descriptive study [(1a) in Table 5.5]. If the expected payoffs are the same, then the expected value of information is the same for prescriptive and descriptive studies (1b). An example in which the prescriptive and descriptive approaches agree for the without-forecast condition is a case study of the fallowing/planting problem (Brown, Katz, and Murphy, 1986; Katz, Brown, and Murphy, 1987; and Chapter 4 of this volume). For the case of climatological information alone, the normative model prescribes an optimal policy of fallowing in alternate years, in agreement with observed behavior. The final logical possibility depicted in the first row of Table 5.5 cannot occur because it requires an expected payoff from the descriptive study to exceed the corresponding expected payoff from the prescriptive studies. As indicated above, this cannot occur if the two models are based on the same information.

(ii) *The prescriptive study yields a lower expected payoff than the descriptive study for the with-forecast decision.* This result would suggest that nonoptimal use of information is limiting its value to the user. Training or implementation of aids designed to help the user optimize information processing could be helpful and might increase the value of the forecast. Depending on the expected payoff of the decision without the forecast, the expected value of information obtained from the descriptive study could be greater than (2a), equal to (2b), or less than (2c) the corresponding value from the prescriptive study (see Table 5.5). Practically speaking, however, the important implication would be the same in all three cases: The value of information would be greater if users made better use of it.

Davis and Nnaji (1982) discuss the possibility that the descriptive study yields a negative estimate (before the cost of the forecast is subtracted). They state that it is most likely to occur when the information is quite uncertain and it is being used nonoptimally. They also state that "presently the negative value of information is more likely to be detected by observation than by analysis and is more likely to be corrected by obtaining more information than by utilizing a better decision rule" (p. 469). They provide no support for this statement. If both prescriptive and descriptive models are available, then it is possible to determine analytically to what extent "utilizing a better decision rule" will correct the situation. At the least, a better decision rule — that is, ignoring the information — can increase the value to zero.

Of course, differences between descriptive and prescriptive models may indicate that either or both of the models are incorrect. For example, either model might include incorrect assumptions about the constraints and limitations that decision makers face or the information that is available to them. Differences between models could serve to highlight areas that need further development and study.

Despite the potential for improved value estimates and insight into the use and value of weather forecasts from combined prescriptive/descriptive studies, no such studies have been conducted. A major reason for this is the lack of descriptive studies that have produced decision models that are sufficiently well-developed to be used to produce a forecast value comparable to that produced by prescriptive techniques. Methods for developing such descriptive models exist, and their application to the study of the use and value of weather forecasts should be encouraged. Such application should prove useful to the meteorological community by providing a more complete understanding of the factors that determine the value of forecasts.

Acknowledgments

The author is grateful to Joel Curtis and Robert Balfour for providing valuable source material for this chapter. Preparation of this chapter was supported by the National Science Foundation under grant No. SES-9109594.

References

Aber, P.G. (1990). Social and economic benefits of weather services: Assessment methods, results, and applications. In *Economic and Social Benefits of Meteorological and Hydrological Services, Proceedings of the Technical Conference*, WMO No. 733, 48–65. Geneva: World Meteorological Organization.

Adams, R.L.A. (1974). The differential use of personal observations and weather forecasts in making New England beach trip decisions. *Preprints, Fifth Conference on Weather Forecasting and Analysis*, 40–43. Boston: American Meteorological Society.

Arkes, H.R. & Hammond, K.R., ed. (1986). *Judgment and Decision Making: An Interdisciplinary Reader*. Cambridge: Cambridge University Press.

Baker, J. (1984). Public response to hurricane probability forecasts. Report, National Weather Service, Weather Analysis and Prediction Division, (NTIS PB84-15868). Silver Spring, MD: National Weather Service.

Baron, J. (1988). *Thinking and Deciding*. New York: Cambridge University Press.

Brand, S. (1992). Applying weather analyses and forecasts in the Navy decision-making process. *Bulletin of the American Meteorological Society*, **73**, 31–33.

Brehmer, A. & Brehmer, B. (1988). What have we learned about human judgment from thirty years of policy capturing? In *Human Judgment: The Social Judgment Theory View*, ed. B. Brehmer & C.R.B. Joyce, 75–114. Amsterdam: North-Holland.

Brehmer, B. (1990). Strategies in real-time, dynamic decision making. In *Insights in Decision Making: A Tribute to Hillel J. Einhorn*, ed. R.M. Hogarth, 262–279. Chicago: University of Chicago Press.

Brehmer, B. & Joyce, C.R.B., ed. (1988). *Human Judgment: The Social Judgment Theory View*. Amsterdam: North-Holland.

Brown, B.G., Katz, R.W. & Murphy, A.H. (1986). On the economic value of seasonal-precipitation forecasts: the fallowing/planting problem. *Bulletin of the American Meteorological Society*, **67**, 833–841.

Brown, B.G. & Murphy, A.H. (1987). The potential value of climate forecasts to the natural gas industry in the United States. Final Report. Gas Research Institute. Chicago, IL.

Brunswik, E. (1956). *Perception and the Representative Design of Psychological Experiments* (second edition). Berkeley: University of California Press.

Changnon, S.A. (1992). Contents of climate predictions desired by agricultural decision makers. *Journal of Applied Meteorology*, **31**, 1488–1491.

Changnon, S.A., Changnon, J.M. & Changnon, D. (1995). Uses and applications of climate forecasts for power utilities. *Bulletin of the American Meteorological Society*, **76**, 711–720.

Changnon, S.A. & Vonnhame, D.R. (1986). Use of climate predictions to decide a water management problem. *Water Resources Bulletin*, **22**, 649–652.

Curtis, J.C. & Murphy, A.H. (1985). Public interpretation and understanding of forecast terminology: some results of a newspaper survey in Seattle, Washington. *Bulletin of the American Meteorological Society*, **66**, 810–819.

Curtis, J.C. & Sites, W.E. (1987). Impact of marine weather warnings on the summertime recreational fishery for the Strait of Juan de Fuca. Unpublished paper presented at the Fifth Symposium on Coastal and Ocean Management. Boston: American Meteorological Society.

Davis, D.R. & Nnaji, S. (1982). The information needed to evaluate the worth of uncertain information, predictions and forecasts. *Journal of Applied Meteorology*, **21**, 461–470.

Dawes, R.M. (1988). *Rational Choice in an Uncertain World.* New York: Harcourt, Brace, Jovanovich.

Del Greco, J. (1983). New Jersey Sea Grant fishing industry study: influence of weather and sea state. New Jersey Sea Grant Publication No. NJSG-83-119, South Hackensack, NJ.

DPA Group, Inc. (1985). The economic value of weather information in Canada. Final report, Atmospheric Environment Service, Environment Canada, Montreal.

Easterling, W.E. (1986). Subscribers to the NOAA *Monthly and Seasonal Weather Outlook. Bulletin of the American Meteorological Society*, **67**, 402–410.

Ebbesen, E.B. & Konecni, V.J. (1980). On the external validity of decision-making research: what do we know about decisions in the real world? In *Cognitive Processes in Choice and Decision Behavior*, ed. S. Wallsten, 21–45. Hillsdale, NJ: Erlbaum.

Einhorn, H.J., Kleinmuntz, D.N. & Kleinmuntz, B. (1979). Linear regression and processing-tracing models of judgment. *Psychological Review*, **86**, 465–485.

Ericsson, K.A. & Simon, H.A. (1984). *Protocol Analysis: Verbal Reports as Data.* Cambridge, MA: MIT Press.

Ewalt, R.E., Wiersma, D. & Miller, W.L. (1973). Operational value of weather information in relation to soil management characteristics. *Agronomy Journal*, **65**, 437–439.

Fischhoff, B. & Furby, L. (1988). Measuring values: a conceptual framework for interpreting transactions with special reference to contingent valuations of visibility. *Journal of Risk and Uncertainty*, **1**, 147–184.

Fraser, J.M., Smith, P.J. & Smith, J.W. (1992). A catalog of errors. *International Journal of Man-Machine Studies*, **37**, 265–307.

Gigerenzer, G. & Murray, D.J. (1987). *Cognition as Intuitive Statistics.* Hillsdale, NJ: Erlbaum.

Glantz, M.H. (1977). The value of a long-range weather forecast for the West African Sahel. *Bulletin of the American Meteorological Society*, **58**, 150–158.

Glantz, M.H. (1980). Considerations of the societal value of an El Niño forecast and the 1972-1973 El Niño. In *Resource Management and Environmental Uncertainty*, ed. M.H. Glantz, 449–476. New York: Wiley.

Glantz, M.H. (1982). Consequences and responsibilities in drought forecasting: the case of Yakima, 1977. *Water Resources Research*, **18**, 3–13.

Hammond, K.R. (1955). Probabilistic functioning and the clinical method. *Psychological Review*, **62**, 255–262.

Hammond, K.R. (1990). Intuitive and analytical cognition: information models. In *Concise Encyclopedia of Information Processing in Systems and Organizations*, ed. A. Sage, 306–312. Oxford: Pergamon Press.

Hammond, K.R., Hamm, R.M., Grassia, J. & Pearson, T. (1987). Direct comparison of the efficacy of intuitive and analytical cognition in expert judgment. *IEEE Transactions on Systems, Man, and Cybernetics*, **SMC-17**, 753–770.

Hammond, K.R., McClelland, G.H. & Mumpower, J. (1980). *Human Judgment and Decision Making: Theories, Methods, and Procedures.* New York: Praeger.

Hammond, K.R., Stewart, T.R., Brehmer, B. & Steinman, D.O. (1975). Social judgment theory. In *Human Judgment and Decision Processes*, ed. M.F. Kaplan & S. Schwartz, 271–312. New York: Academic Press.

Hawando, T. (1990). Application of climatic data in soil resource management for increased and sustainable agricultural production: a case study from Ethiopia. In *Economic and Social Benefits of Meteorological and Hydrological Services, Proceedings of the Technical Conference*, WMO No. 733, 171–182. Geneva: World Meteorological Organization.

Herath, H.M.G., Hardaker, J.B. & Anderson, J.R. (1982). Choice of varieties by Sri Lanka rice farmers: comparing alternative decision models. *American Journal of Agricultural Economics*, **64**, 87–93.

Hoffman, P.J. (1960). The paramorphic representation of clinical judgment. *Psychological Bulletin*, **57**, 116–131.

Hofing, S.L., Sonka, S.T. & Changnon, S.A. (1987). Enhancing information use in decision making: agribusiness and climate information. Final report, NSF IS 86-60497. Champaign, IL: Agricultural Education and Consulting.

Hogarth, R.M. (1981). Beyond discrete biases: functional and dysfunctional aspects of judgmental heuristics. *Psychological Bulletin*, **90**, 197–217.

Hogarth, R. M. (1987). *Judgement and Choice: The Psychology of Decision.* Chichester, UK: Wiley.

Janis, I.L. & Mann, L. (1977). *Decision Making: A Psychological Analysis of Conflict, Choice, and Commitment.* New York: Free Press.

Johnson, S.R. (1990). Practical approaches for uses of economic principles in assessing the benefits of meteorological and hydrological services. In *Economic and Social Benefits of Meteorological and Hydrological Services, Proceedings of the Technical Conference*, WMO No. 733, 12–33. Geneva: World Meteorological Organization.

Kahneman, D., Slovic, P. & Tversky, A. (1982). *Judgment Uncertainty, Heuristics and Biases.* Cambridge: Cambridge University Press.

Katz, R.W., Brown, B.G. & Murphy, A.H. (1987). Decision-analytic assessment of the economic value of weather forecasts: the fallowing/planting problem. *Journal of Forecasting*, **6**, 77–89.

Katz, R.W., Murphy, A.H. & Winkler, R.L. (1982). Assessing the value of frost forecasts to orchardists: a dynamic decision-making approach. *Journal of Applied Meteorology*, **21**, 518–531.

Keeney, R.L. & Raiffa, H. (1976). *Decisions with Multiple Objectives: Preferences and Value Tradeoffs.* New York: Wiley.

Kleinmuntz, B., ed. (1968). *Formal Representation of Human Judgment.* New York: Wiley.

Kruglanski, A.W., Friedland, N. & Farkash, E. (1984). Lay persons' sensitivity to statistical information: the case of high perceived applicability. *Journal of Personality and Social Psychology*, **46**, 503–518.

Lave, L.B. (1963). The value of better weather information to the raisin industry. *Econometrica*, **31**, 151–164.

Lopes, L.L. (1991). The rhetoric of rationality. *Theory & Psychology*, **1**, 65–82.

Lusk, C.M., Stewart, T.R., Hammond, K.R. & Potts, R.J. (1990). Judgment and decision making in dynamic tasks: the case of forecasting the microburst. *Weather and Forecasting*, **5**, 627–639.

Mathews, J.H. (1992). The art of terminal forecasting. *Air Traffic Bulletin*, FAA, No. 92-1, 4–7.

McNew, K.P., Mapp, H.P., Duchon, C.E. & Merritt, E.S. (1991). Sources and uses of weather information for agricultural decision makers. *Bulletin of the American Meteorological Society*, **72**, 491–498.

McQuigg, J.D. (1971). Some attempts to estimate the economic response of weather information. *Weather*, **26**, 60–68.

Mitchell, R.C. & Carson, R.T. (1989). *Using Surveys to Value Public Goods: The Contingent Valuation Method.* Washington, D.C.: Resources for the Future.

Mjelde, J.W., Dixon, B.L. & Sonka, S.T. (1989a). Estimating the value of sequential updating solutions for intrayear crop management. *Western Journal of Agricultural Economics*, **14**, 1–8.

Mjelde, J.W. & Frerich, S.J. (1987). Selected review of literature concerned with socioeconomic issues of climate/weather forecasting with additional references. Departmental Information Report DIR 87-1/SP-5, The Texas Agricultural Experiment Station, Texas A&M University, College Station, TX.

Mjelde, J.W., Sonka, S.T. & Peel, D.S. (1989b). The socioeconomic value of climate and weather forecasting: a review. Research Report 89-01, Midwestern Climate Center, Illinois State Water Survey, Champaign, IL.

Montgomery, H. (1984). Decision rules and the search for dominance structure: towards a process model of decision making. In *Analyzing and Aiding Decision Processes*, ed. P.C. Humphreys, O. Svenson & A. Vari, 343–369. Amsterdam: North-Holland.

Mroz, P.J. & Raven, R.J. (1993). Levels of student understanding and reasoning associated with televised weather information. *Bulletin of the American Meteorological Society*, **74**, 425–438.

Murphy, A.H. & Brown, B.G. (1982). User requirements for very-short-range weather forecasts. In *Nowcasting*, ed. K.A. Browning, 3–15. New York: Academic Press.

Murphy, A.H. & Brown, B.G. (1983). Forecast terminology: composition and interpretation of public weather forecasts. *Bulletin of the American Meteorological Society*, **64**, 13–22.

Murphy, A.H., Lichtenstein, S., Fischhoff, B. & Winkler, R.L. (1980). Misinterpretations of precipitation probability forecasts. *Bulletin of the American Meteorological Society*, **61**, 695–701.

Nisbett, R.E., Krantz, D.H., Jepson, C. & Kunda, Z. (1983). The use of statistical heuristics in everyday inductive reasoning. *Psychological Review*, **90**, 339–363.

Oliver, R.M. & Smith, J.Q., ed. (1990). *Influence Diagrams, Belief Nets, and Decision Analysis*. New York: Wiley.

Payne, J.W., Bettman, J.R. & Johnson, E.J. (1992). Behavioral decision research: a constructive processing perspective. *Annual Review of Psychology*, **43**, 87–131.

Prototype Regional Observing and Forecasting Service (1979). Report of a study to estimate economic and convenience benefits of improved local weather forecasts. NOAA Technical Memorandum ERL PROFS-1. Boulder, CO: NOAA Environmental Research Laboratory.

Roebber, P.J. & Bosart, L.F. (1996). The complex relationship between forecast skill and forecast value: a real-world analysis. *Weather and Forecasting*, **11**, 544–559.

Ryder, P. (1990). The assessment and testing of user requirements for specific weather and climate services. In *Economic and Social Benefits of Meteorological and Hydrological Services, Proceedings of the Technical Conference*, WMO No. 733, 103–107. Geneva: World Meteorological Organization.

Sonka, S.T., Changnon, S.A. & Hofing, S. (1988). Assessing climate information use in agribusiness. II: decision experiments to estimate economic value. *Journal of Climate*, **1**, 766–774.

Sonka, S.T., Changnon, S.A. & Hofing, S. (1992). How agribusiness uses climate predictions: implications for climate research and provision of predictions. *Bulletin of the American Meteorological Society*, **73**, 1999–2008.

Sterman, J.D. (1989). Modeling managerial behavior: misperceptions of feedback in a dynamic decision making experiment. *Management Science*, **35**, 321–339.

Stewart, T.R. (1988). Judgment analysis: procedures. In *Human Judgment: The Social Judgment Theory View*, ed. B. Brehmer & C.R.B. Joyce, 41–74. Amsterdam: North-Holland.

Stewart, T.R., Katz, R.W. & Murphy, A.H. (1984). Value of weather information: a descriptive study of the fruit-frost problem. *Bulletin of the American Meteorological Society*, **65**, 126–137.

Stewart, T.R., Moninger, W.R., Grassia, J., Brady, R.H. & Merrem, F.H. (1989). Analysis of expert judgment in a hail forecasting experiment. *Weather and Forecasting*, 4, 24–34.

Suchman, D., Auvine, B. & Hinton, B. (1979). Some economic effects of private meteorological forecasting. *Bulletin of the American Meteorological Society*, **60**, 1148–1156.

Suchman, D., Auvine, B. & Hinton, B. (1981). Determining the economic benefits of satellite data in short-range forecasting. *Bulletin of the American Meteorological Society*, **62**, 1458–1465.

Tversky, A. & Kahneman, D. (1974). Judgment under uncertainty: heuristics and biases. *Science*, **185**, 1124–1131.

Tversky, A. & Kahneman, D. (1981). The framing of decisions and the rationality of choice. *Science*, **211**, 453–458.

Wallace, H.A. (1923). What is in the corn judge's mind? *Journal of the American Society of Agronomy*, **15**, 300–304.

6

Forecast value:
prototype decision-making models

RICHARD W. KATZ
and
ALLAN H. MURPHY

1. Introduction

In this chapter, theoretical relationships between the scientific quality and economic value of imperfect weather forecasts are considered. Prototype decision-making models are treated that, while relatively simple in structure, still capture some of the essential features (e.g., the "dynamics") of real-world situations. The emphasis is on analytical results that can be obtained concerning the structure of the optimal policy and the shape of the "quality/value curve." It is anticipated that knowledge of such theoretical properties for prototype models will provide insight into analogous properties for real-world decision-making problems that are inherently much more complex.

The prospects for increases in the quality of weather forecasts in the future have been discussed in Chapter 1 of this volume, thereby providing a partial justification for the hypothetical increases in forecast quality that are assumed in the present chapter. Various aspects of forecast quality (e.g., bias and accuracy) have been described in Chapter 2. A simplified form of weather information is treated here in which a one-dimensional measure of quality can be defined that is synonymous with the concept of sufficiency (also described in Chapter 2). The Bayesian decision-theoretic approach to assessing the economic value of imperfect information is adopted, an approach that has been introduced in Chapter 3. Analytical results obtained for prototype decision-making models are compared with those for real-world case studies that have also employed this prescriptive approach to real-world decision-making problems (reviewed in Chapter 4). Reasons for the underutilization (or nonoptimal use) of existing weather forecasts have been outlined in Chapter 5, and some explanations for such behavior based on theoretical considerations are presented here.

In Section 2, first the concepts of scientific quality, sufficiency, economic value, and optimal policy in a general Bayesian decision-analytic setting are briefly reviewed. The few relationships among these concepts that always hold are identified. Then the prototype form of weather information on which the present chapter focuses is introduced. Only two states of weather are allowed, and only two conditional probabilities are required to characterize fully the quality of the imperfect weather forecasting system. Several forms of prototype decision-making models are described in Section 3, including the static cost–loss ratio (or "umbrella") problem and generalizations to dynamic situations, that treat both finite-horizon, undiscounted and infinite-horizon, discounted problems. For each model, the structure of the optimal policy and the shape of the quality/value curve can be expressed in closed form. In an attempt to broaden the scope, extensions of these prototype decision-making models are considered in Section 4. These extensions primarily involve allowing for more complex, and consequently more realistic, forms of information about weather. Finally, Section 5 deals with the implications of these results, including their relevance for real-world decision-making problems. Some technical results are relegated to an appendix. We note that Ehrendorfer and Murphy (1992b) have also reviewed quality/value relationships for imperfect weather forecasting systems.

2. Concepts

2.1. General setting

As defined in Chapter 2 of this volume, the scientific quality of a forecasting system is the totality of the statistical characteristics embodied in the joint distribution of forecasts and observations (equivalent definitions can be formulated in terms of conditional and marginal distributions). In general, more than one number is required to describe forecast quality completely; that is, forecast quality is inherently a multidimensional concept. Thus, scalar measures of the correspondence between forecasts and observations, such as the mean square error or a correlation coefficient, represent measures of particular aspects of quality (i.e., accuracy or linear association). These measures are usually based solely on the forecasts and observations, and they do not make explicit use of any of the economic parameters associated with a specific decision-making problem. Typically, such measures can be scaled so that they range from zero for a forecasting system with no predictive ability to one for perfect information, with the score for imperfect forecasts falling between these two limits. The relative merits of common measures of various aspects of forecast quality have been discussed in Chapter 2.

The Bayesian decision-analytic concept of the *economic value* of imperfect weather forecasts is predicated upon the existence of an individual decision maker who considers making use of the forecasts. For a particular decision-making problem, this individual is assumed to select the action that maximizes his (or her) expected utility. Here the expectation is taken with respect to the information available about the future weather. A rule that specifies which action the decision maker should take as a function of the information received is termed the *optimal policy*. This information could be in the form of a genuine forecast. On the other hand, even if no such forecast were available, it is reasonable to assume that the decision maker would still be aware of the long-run statistical behavior of the weather variable (termed "climatological information"). Economic value is thus measured by comparing the expected utilities with and without the forecast. This concept of information value has been described in more detail in Chapter 3 of this volume.

Measures of aspects of quality do not necessarily have any straightforward connection to the economic value of weather forecasts. In fact, it is possible for one forecasting system to have higher accuracy (according to some reasonable one-dimensional measure) than another system, and yet the inferior system (in terms of accuracy) still produces forecasts of higher economic value for certain decision-making problems or decision makers (e.g., Murphy and Ehrendorfer, 1987). Another concept, namely *sufficiency*, needs to be introduced in order to produce any consistent connection between the scientific quality of forecasts and their economic value in general. Essentially, one forecasting system is sufficient for a second one, if the forecast information generated by the second can be obtained by a stochastic transformation of the information generated by the first system. Here the forecast information is characterized by the conditional distribution of forecasts given the weather observation, and the stochastic transformation involves a conditioning (or "nesting") operation, termed the "sufficiency relation." The precise definition of sufficiency has been given in Chapter 2 of this volume (also see Ehrendorfer and Murphy, 1988, 1992a; Krzysztofowicz and Long, 1991). Here the important point to note is that if one forecasting system is sufficient for another, then it is guaranteed to result in at least as much economic value as the other system, no matter what the decision maker's loss or payoff function (Blackwell, 1953).

For some special forms of weather information, it is possible to define a one-dimensional measure that satisfies the sufficiency relation, and thereby measures forecast quality in its entirety. In this case, for any given decision-making problem, economic value must be a nondecreasing function of quality. However, nothing can still be said in general about the rate at which such a quality/value curve increases (e.g., is it convex or concave?). Moreover, the sufficiency concept provides just about the only instance in which a change in one of the attributes of a decision-making problem results in a monotonic relationship with economic value (Hilton, 1981).

2.2. Prototype form of weather information

A simplified form of weather information is treated in the remainder of the present chapter. The weather variable, a random variable, denoted by Θ, has only two possible states:

(i) *adverse weather* $(\Theta = 1)$;
(ii) *no adverse weather* $(\Theta = 0)$.

Climatological information consists of a single probability of adverse weather

$$p_\Theta = \Pr\{\Theta = 1\}, \qquad (6.1)$$

perhaps derived from historical weather records. From a Bayesian perspective, the parameter p_Θ can be viewed as the "prior probability" of adverse weather.

Imperfect information about Θ consists of a random variable Z, indicating a forecast of adverse weather $(Z = 1)$ or of no adverse weather $(Z = 0)$. This simple forecasting system is completely characterized by the two conditional probabilities of adverse weather

$$p_1 = \Pr\{\Theta = 1 | Z = 1\}, \quad p_0 = \Pr\{\Theta = 1 | Z = 0\}. \qquad (6.2)$$

Without loss in generality, it is assumed that these conditional probabilities satisfy the ordering $0 \leq p_0 \leq p_\Theta \leq p_1 \leq 1$. This parameter configuration is shown in Figure 6.1a. The limiting circumstance of no forecasting ability (i.e., climatological information) corresponds to $p_0 = p_1 = p_\Theta$, whereas the other extreme of perfect information corresponds to $p_0 = 0$ and $p_1 = 1$. This form of imperfect forecasts has been termed "categorical," when viewed from an *ex post* perspective, whereas it is a "primitive" case of probabilistic forecasts in which only two possible probabilities of adverse weather are ever issued when viewed from an *ex ante* perspective (see Chapter 3 of this volume for an explanation of the terms *ex post* and *ex ante*). In Bayesian terminology, the parameters, p_0 and p_1, would be termed "posterior probabilities" of adverse weather.

For this prototype form of weather information, the sufficiency relation reduces to a very simple condition on the two conditional probabilities of the forecasting system. Consider another forecasting system with possibly different conditional probabilities, denoted by p'_0 and p'_1. Then the original forecasting system is *sufficient* for the other system if and only if

$$p_0 \leq p'_0 \leq p_\Theta \leq p'_1 \leq p_1. \tag{6.3}$$

This condition (6.3) is illustrated in Figure 6.1b, and has appeared either in this form or in an equivalent form in DeGroot (1970, p. 444), Ehrendorfer and Murphy (1988), and Krzysztofowicz and Long (1990). In order to guarantee that one forecasting system produces at least as great an economic value for all decision makers as the other no matter what the decision makers' loss or payoff functions, both conditional probabilities must be at least as far away from the climatological probability p_Θ as the corresponding ones for the other system, an intuitively appealing requirement. The intermediate case (i.e., either $p_0 \leq p'_0$ and $p_1 \leq p'_1$, or $p'_0 \leq p_0$ and $p'_1 \leq p_1$) is indeterminate with respect to economic value (see Figure 6.1c). Specifically, for some loss or payoff functions the first system will be superior to the second in terms of economic value, whereas for other loss or payoff functions the second system will be superior to the first (e.g., Murphy and Ehrendorfer, 1987).

To simplify the characterization of the forecasting system further, one additional requirement will be imposed. The plausible assumption is made that forecasts of adverse weather are issued with the same long-run relative frequency as the occurrence of

(a)

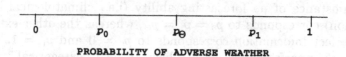

PROBABILITY OF ADVERSE WEATHER

(b)

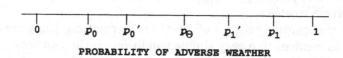

PROBABILITY OF ADVERSE WEATHER

(c)

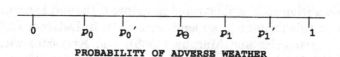

PROBABILITY OF ADVERSE WEATHER

adverse weather (termed "overall reliability" or "unconditionally unbiased"); that is,

$$\Pr\{Z = 1\} = p_\Theta. \tag{6.4}$$

This requirement is equivalent to constraining the conditional probability p_0 to move from p_Θ toward zero at the same relative rate at which the conditional probability p_1 moves from p_Θ toward one (see Figure 6.1d). In particular, equation (6.4) implies that

$$p_0 = \frac{(1 - p_1)p_\Theta}{1 - p_\Theta}. \tag{6.5}$$

That is, p_0 is simply $1 - p_1$ weighted by the climatological "odds" of adverse weather, $p_\Theta/(1 - p_\Theta)$. The forecasting system is then completely characterized by either one of the two conditonal probabilities in equation (6.2), say p_1, for a fixed climatological prob-

(d)

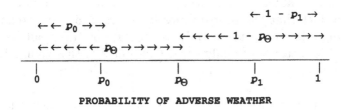

PROBABILITY OF ADVERSE WEATHER

(e)

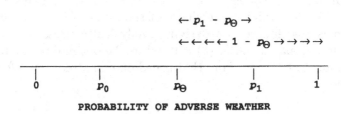

PROBABILITY OF ADVERSE WEATHER

Figure 6.1. Relationship among parameters of imperfect weather forecasting system: (a) basic parameter configuration; (b) example in which sufficiency condition (6.3) is satisfied; (c) example in which sufficiency condition (6.3) is *not* satisfied; (d) constraint (6.5) imposed by requirement of overall reliability, equation (6.4); (e) measure of forecast quality, equation (6.6).

ability of adverse weather p_Θ. This condition (6.5) of overall reliability of the forecasts means that the sufficiency relation (6.3) is automatically satisfied as p_1 is increased between p_Θ and one.

It is convenient to rescale the conditional probability of adverse weather p_1, defining the measure of *quality* for this prototype form of imperfect weather information as

$$q = \frac{p_1 - p_\Theta}{1 - p_\Theta}, \tag{6.6}$$

just the relative distance of p_1 between p_Θ and one (see Figure
6.1e). Note that $0 \le q \le 1$, with $q = 0$ for climatological information and $q = 1$ for perfect information. It is easy to show that q
is equal to the ordinary correlation between the forecast variable
Z and the weather variable Θ [i.e., $q = \mathrm{Corr}(Z, \Theta)$, where "Corr"
denotes the coefficient of correlation]. Because the unconditional
and conditional biases have been assumed away (i.e., condition 6.4
and the fact that the probabilities in equation 6.2 are "reliable"
in an *ex ante* sense), the correlation coefficient turns out to be a
reasonable measure of quality (see Chapter 2 of this volume).

Finally, it is important to reiterate the implications for the economic value of the imperfect weather forecasting system of the
fact that the quality measure q satisfies the sufficiency relation.
No matter what the particular form of decision maker's loss or
payoff function, the economic value, denoted by $V(q)$, must be a
nondecreasing function of the quality q of the forecasting system.
The graph of $V(q), 0 \le q \le 1$, is referred to as the *quality/value
curve* [note, by definition, $V(0) = 0$]. What remains to be done
is to obtain analytical expressions for $V(q)$ for certain prototype
decision-making models. Because the optimal policy plays a cru-
cial role in determining the shape of the quality/value curve, attention is also focused on deriving analytical expressions for its
structure. In particular, $V(q) = 0$ unless the optimal policy based
on the forecasts differs from that based on climatological information alone.

3. Prototype decision-making models

All of the prototype decision-making models treated in this section
are versions of the so-called cost–loss ratio model. This problem
has essentially the simplest possible form of decision maker's loss
or payoff function. Nevertheless, it has a long tradition of study,
especially within the meteorological community. The cost–loss ratio model was apparently introduced by Thompson (1952). Note
that, in this chapter, the term "cost–loss" refers to a more specific class of decision-making models than does the same term in
Chapter 3 of this volume.

3.1. Static cost–loss ratio model

The decision maker must choose between two possible actions:

Table 6.1. Expense matrix for cost–lost ratio decision-making model

Action	Weather state	
	Adverse $(\Theta = 1)$	Not adverse $(\Theta = 0)$
Protect	C	C
Do not protect	L	0

(i) *protect*;
(ii) *do not protect*.

If protective action is taken, then the decision maker incurs a
cost C, no matter what the weather. If protective action is not
taken and adverse weather does occur (i.e., $\Theta = 1$), then the
decision maker incurs a *loss L*, $0 < C < L$. The complete expense
matrix for this decision-making problem is given in Table 6.1. The
so-called umbrella problem refers to the situation in which the
protective action is to "take an umbrella" and the adverse weather
is "rain" (Katz and Murphy, 1987).

For now, the static version of the cost–loss ratio model is considered. This situation could be viewed as a "one-shot" problem,
in which the decision maker need only choose an action to take
on a single occasion. On the other hand, it could be viewed as a
repetitive problem in which the decision maker faces the identical situation on a sequence of occasions, with no carryover effects
from one occasion to the next (i.e., no "dynamics"). In particular, the fact that a loss L has been incurred on one occasion does
not preclude the decision maker suffering the same loss again on
some future occasion. Further, the state of weather on one occasion is taken, for the time being, to be independent of that on any
subsequent occasion.

The decision maker must select the action (either protect or do
not protect) that achieves the desired goal of minimizing the expected expense (sometimes termed "Bayes risk"). This criterion is
equivalent to maximizing expected return, which is in turn a special case of maximizing expected utility when the decision maker's
utility function is linear in monetary expense (see Chapter 3 of this
volume). The economic value of the imperfect weather forecasting system is the reduction in expected expense associated with

the forecasts, as compared to the expected expense when only the climatological probability of adverse weather, equation (6.1), is available. Formally, the minimal expected expense, expressed as a function of the quality q of the weather forecasting system, equation (6.6) in Section 2.2, is denoted by $E(q)$. The corresponding economic value of the forecasts is

$$V(q) = E(0) - E(q), \quad 0 \leq q \leq 1. \tag{6.7}$$

As stated in Section 2.2, $V(q)$ must be a nondecreasing function of q. For the static cost–loss ratio problem, a simple analytical expression for this quality/value curve can be obtained.

Example [Climatological information (i.e., $q = 0$) with $C = 0.25$ and $L = 1$]. When protective action is taken, the decision maker incurs a cost $C = 0.25$, no matter what the climatological probability of adverse weather. If this probability $p_\Theta = 0.3$, for example, then the expected expense when protective action is not taken is $p_\Theta L = (0.3)(1) = 0.3$. Because $0.25 < 0.3$, the decision maker should take protective action. On the other hand, if $p_\Theta = 0.2$, for example, then $p_\Theta L = (0.2)(1) = 0.2 < 0.25$ and the decision maker should not take protective action. Evidently, for fixed economic parameters C and L, whether the decision maker should take protective action is governed by the likelihood of occurrence of adverse weather.

This example serves to motivate the following general results for the case of climatological information alone. The structure of the optimal policy is to take protective action provided $C < p_\Theta L$; that is,

(i) *protect if $p_\Theta > C/L$;*
(ii) *do not protect if $p_\Theta < C/L$.*

The fact that the optimal policy depends on the two economic parameters only through the ratio C/L explains the origin of the name "cost–loss ratio" decision-making situation (e.g., Murphy, 1977). The corresponding minimal expected expense for climatological information is

$$E(0) = \min\{C, \ p_\Theta L\}. \tag{6.8}$$

This expected expense expression provides one of the two inputs required in equation (6.7) to determine the economic value of an imperfect weather forecasting system.

Example (Imperfect forecasts with $C = 0.25$, $L = 1$, and $p_\ominus = 0.2$).
It has already been established that this probability of adverse
weather, $p_\ominus = 0.2$, is too low for protection to be optimal with
climatological information alone. But suppose that an imperfect
weather forecasting system exists, for example, with quality $q = 0.05$ (i.e., the conditional probabilities of adverse weather are $p_1 = 0.24$ and $p_0 = 0.19$ from equations 6.5 and 6.6). According to
assumption (6.4), the likelihood that a forecast of $Z = 1$ is issued
is $p_\ominus = 0.2$ (i.e., the conditional probability of adverse weather is
$p_1 = 0.24$). The likelihood that a forecast of $Z = 0$ is issued is
$1 - p_\ominus = 0.8$ (i.e., the conditional probability of adverse weather
is $p_0 = 0.19$). First, suppose that the decision maker receives a
forecast of $Z = 1$. If protective action is taken, then the expected
expense is again $C = 0.25$. If protective action is not taken, the
expected expense is now $p_1 L = (0.24)(1) = 0.24$. Because $0.24 < 0.25$, the decision maker should still not take protective action.
Second, suppose that $Z = 0$. The expected expense if protective
action is not taken is $p_0 L = (0.19)(1) = 0.19 < 0.25$. So the
optimal policy remains never to take protective action, in spite
of the availability of the forecasts. Consequently, forecasts of this
particular quality level are of no economic value for this specific
decision maker [as a check, $E(0.05) = (0.2)(0.24) + (0.8)(0.19) = 0.2 = E(0)$].

On the other hand, suppose that a weather forecasting system
exists with a higher quality, for example, $q = 0.5$ (i.e., $p_1 = 0.6$
and $p_0 = 0.1$). If $Z = 1$, then the expected expense when protective action is not taken increases to $p_1 L = (0.6)(1) = 0.6 > 0.25$,
and protective action should be taken. If $Z = 0$, then this expected expense is $p_0 L = (0.1)(1) = 0.1 < 0.25$, and protective action should not be taken. So, weighting by the likelihood
of each possible forecast being issued, the minimal expected expense associated with this imperfect weather forecasting system is
$E(0.5) = p_\ominus C + (1 - p_\ominus)(p_0 L) = (0.2)(0.25) + (0.8)(0.1) = 0.13$.
The economic value of the forecasts is $V(0.5) = E(0) - E(0.5) = 0.2 - 0.13 = 0.07$. In other words, the decision maker would save
an average of 7/100th of the possible loss L by relying on the forecasting system instead of climatological information. Evidently,
the quality must exceed a certain threshold for the weather forecasting system to be of any economic value to the decision maker.

Again, this example motivates the following general results for
the imperfect weather forecasting system. First, as in the example,

suppose that $0 < p_\Theta < C/L$. Provided $0 \le q < q^*$, where

$$q^* = \frac{C/L - p_\Theta}{1 - p_\Theta} \tag{6.9}$$

(i.e., $p_\Theta \le p_1 < C/L$), it is never optimal to protect no matter what the weather forecast Z. The minimal expected expense is $E(q) = E(0) = p_\Theta L$, and $V(q) = 0, 0 \le q < q^*$. On the other hand, provided $q^* < q \le 1$ (i.e., $C/L < p_1 \le 1$), the optimal policy is of the form

   (i) *protect* if $Z = 1$;
   (ii) *do not protect* if $Z = 0$.

Using equations (6.5) and (6.6), the minimal expected expense is

$$\begin{aligned} E(q) &= p_\Theta C + (1 - p_\Theta)p_0 L = p_\Theta[C + (1 - p_1)L] \\ &= p_\Theta[C + (1 - p_\Theta)(1 - q)L], \quad q^* \le q \le 1. \end{aligned} \tag{6.10}$$

Using equations (6.7) and (6.10), the economic value of the imperfect weather forecasting system is

$$V(q) = E(0) - E(q) = p_\Theta\{[p_\Theta + (1 - p_\Theta)q]L - C\}, \quad q^* \le q \le 1. \tag{6.11}$$

In other words, the quality/value curve is zero below the quality threshold q^* defined by equation (6.9), and rises linearly above this threshold. Interestingly, the slope of this linear function of q is proportional to the climatological variance, $\text{Var}(\Theta) = p_\Theta(1 - p_\Theta)$, which can be viewed as the "uncertainty" in the underlying weather variable. Figure 6.2 illustrates the piecewise linear shape (in particular, convex — the derivative, provided it exists, of a convex function is itself a nondecreasing function) of this quality/value curve. For the example of $C = 0.3$, $L = 1$, and $p_\Theta = 0.2$, the quality threshold is $q^* = 0.125$ (see equation 6.9) and the slope of the straight line above the threshold is $\text{Var}(\Theta) = 0.16$. Analogous expressions hold for the alternative case in which $C/L < p_\Theta < 1$ (see Katz and Murphy, 1987, for further details).

3.2. Dynamic cost–loss ratio model (finite horizon)

As noted previously, most real-world decision-making problems are dynamic, rather than static, in nature. That is, current and

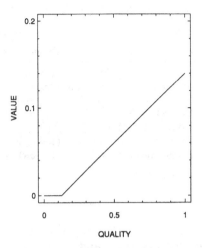

Figure 6.2. Quality/value curve for static cost–loss ratio decision-making model, where $C = 0.3$, $L = 1$, and $p_\Theta = 0.2$. (From Katz and Murphy, 1987)

previous actions and their economic consequences influence future actions and consequences. For instance, the so-called fruit-frost problem is an inherently dynamic decision-making problem with substantial economic ramifications (Baquet, Halter, and Conklin, 1976; Katz, Murphy, and Winkler, 1982). An orchardist must decide whether or not to protect (e.g., by employing heaters, sprinklers, or wind machines) fruit trees during the spring when the buds are especially vulnerable to damage from freezing temperatures. Because the buds damaged or killed cannot recover, it is a dynamic problem. To aid the orchardist in making this decision, each evening during the frost-protection season a minimum temperature forecast is provided. The goal of the orchardist is to minimize the expected expense, not over a single night, but totaled over the entire season (i.e., a finite number of nights) (see Chapter 4 of this volume and Katz et al., 1982 for more details). This type of dynamic decison-making problem is termed "finite horizon."

It is our intent to mimic the dynamic nature of the fruit-frost problem in generalizing the cost–loss ratio model from the static situation, already treated in Section 3.1, to a dynamic situation. Just as buds already killed cannot recover, it is assumed that the loss L can be incurred once at most. The goal of the decision maker is to minimize the expected expense totaled over a finite horizon, with the number of occasions denoted by n. Otherwise, the prob-

Table 6.2. Expected expense for possible strategies for two-stage dynamic cost–loss ratio decision-making model with climatological information

Strategy		Expected expense		
First occasion	Second occasion	First occasion	Second occasion	Total
(i) Protect	Protect	C	C	$2C$
(ii) Protect	Do not protect	C	$p_\Theta L$	$C + p_\Theta L$
(iii) Do not protect	Protect	$p_\Theta L$	$(1 - p_\Theta)C$	$C + p_\Theta(L - C)$
(iv) Do not protect	Do not protect	$p_\Theta L$	$(1 - p_\Theta)p_\Theta L$	$(2 - p_\Theta)p_\Theta L$

lem is unchanged from the static cost–loss ratio model, retaining the same economic parameters, C and L, and the same form of climatological information (with probability of adverse weather p_Θ) and imperfect weather forecasting system (with quality q). Nevertheless, the introduction of this very simplified form of dynamics into the decision-making model results in a much more complex structure of the optimal policy and shape of the quality/value curve. To motivate the dynamic version of the problem, first the simplest case of only a two-occasion (i.e., $n = 2$) situation is considered as an example.

Example ($n = 2$). Table 6.2 lists the total expected expenses for the four possible strategies (i.e., the combinations of either protect or do not protect on the first and second occasions) when climatological information alone is available to the decision maker. To see how the dynamics enters into this problem, we consider the third strategy in this table (i.e., do not protect on the first occasion, but protect on the second occasion). The expected expense on the second occasion is $(1 - p_\Theta)C$, *not* C, because protective action needs to be taken on the second occasion only if the loss L has *not* been incurred on the first occasion (i.e., with probability $1 - p_\Theta$).

As a numerical example, suppose that $C = 0.25$, $L = 1$, and $p_\Theta = 0.3$. Substituting these numerical values into the expressions in Table 6.2, the total expected expenses are:

strategy (i): 0.5;
strategy (ii): 0.55;
strategy (iii): 0.475;

strategy (iv): 0.51.

Strategy (iii), of protecting only on the second occasion, is optimal. Recall that the optimal policy for the static model (i.e., $n = 1$) with the same numerical values of the parameters C, L, and p_Θ, is to take protective action (established in a previous example). If this strategy were naively followed on both occasions [i.e., strategy (i) in Table 6.2], an additional, unnecessary expected expense of $0.5 - 0.475 = 0.025$ (or 1/40th of the possible loss L) would be incurred. The lesson is that following the strategy that minimizes the immediate expected expense on a given occasion does not necessarily correspond to minimizing the total expected expense over all the occasions (two, in this example), because the dynamics of the decision-making model have been ignored.

Returning to the general two-occasion model, it is evident from Table 6.2 that strategy (iii) is always superior to strategy (ii). The motivation for this result is that, given that the decision maker can afford to protect on at most only one of the two occasions, it is preferable to postpone protection until the second occasion, allowing for the possibility that the loss L will be incurred on the first occasion (making protection unnecessary). Upon reaching the second (and last) stage of the decision-making process, it is also evident from this table that the prescribed action is identical to the optimal policy for the static ($n = 1$) problem; namely, protect on the second occasion if $p_\Theta > C/L$. It is only on the first occasion that the prescribed action differs from that for the static situation. Murphy et al. (1985) and Winkler and Murphy (1985) have treated this two-occasion version of the dynamic cost–loss ratio model in more detail.

It should be evident from this example that enumerating all possible strategies for the general n-occasion model with climatological information alone, or even for the two-occasion model when an imperfect forecasting system is available, would be quite tedious. An alternative approach, based on the idea of "backward induction," is suggested by the example just treated. First, the one-occasion (i.e., $n = 1$) or static problem is solved (corresponding to the last occasion). Then, expressing the two-occasion problem as a function of the one-occasion problem and making use of the solution already obtained for the last occasion, a complete solution to the two-occasion problem can be obtained. Continuing to work backward until the first of the n occasions is reached, and expressing the n-occasion problem as a function of the $(n-1)$-occasion problem already solved, produces a solution to the general n-occasion problem. This method of determining the optimal policy on each of the n occasions and the associated minimal total expected expense over the n occasions involves a recursive equation, and is referred to as "stochastic dynamic programming" (e.g., Ross, 1983; White, 1978).

For relatively complex dynamic decision-making models, the recursive equation is employed purely to generate numerical solutions. The use of stochastic dynamic programming for this purpose has been cited in Chapter 4 of this volume. For models whose structure is relatively simple, such as the dynamic cost–loss ratio model, the recursion can be solved analytically. In particular, Murphy et al. (1985) obtained certain analytical results concerning the structure of the optimal policy for the general n-occasion problem, whereas Krzysztofowicz and Long (1990) extended this approach to produce complete analytical solutions for both the structure of the optimal policy and the corresponding minimal total expected expense over the n occasions. White (1966) was apparently the first researcher to advocate the approach of stochastic dynamic programming in assessing the economic value of forecasts, although he did not explicitly consider weather forecasts.

Rather than reproduce these solutions here, we shall be content to describe the essential features of the results. First, a description is provided of how the stochastic dynamic programming recursions for the n-occasion model are derived, expressions from which all the analytical results originate. In keeping with the notation introduced in Section 3.1, let $E_n(q)$, $n = 1, 2, \ldots$, denote the minimal expected expense totaled over the n occasions when the imperfect weather forecasting system with quality q (see Section 2.2) is available to the decision maker. As in equation (6.7), the corresponding economic value of the forecasting system is

$$V_n(q) = E_n(0) - E_n(q), \ 0 \le q \le 1. \tag{6.12}$$

To simplify matters, first consider the situation in which only climatological information is available to the decision maker (i.e., quality $q = 0$). For the one-occasion (or static) cost–loss ratio model, the expression (6.8) already derived in Section 3.1 can be restated with this new notation as

$$E_1(0) = \min\{C, \ p_\Theta L\}. \tag{6.13}$$

For the two-occasion model (i.e., $n = 2$), the minimal total expected expense $E_2(0)$ can be expressed as a function of the minimal expected expense $E_1(0)$ for the static model as follows:

$$E_2(0) = \min\{C + E_1(0),\ p_\Theta L + (1 - p_\Theta)E_1(0)\}. \tag{6.14}$$

The first (second) term on the right-hand side of equation (6.14) represents the total expected expense over the last two occasions when protective action is (is not) taken on the next to last occasion. The first term consists of the cost C of protecting on the first occasion, as well as the minimal expected expense on the second (last) occasion. The second term consists of the loss L on the first occasion, which is incurred with probability p_Θ, as well as the minimal expected expense on the second occasion, which is incurred with probability $1 - p_\Theta$. It is also easy to verify that the expressions already listed for $E_2(0)$ in Table 6.2 satisfy equation (6.14).

Similarly, for the general n-occasion model, the minimal total expected expense with climatological information $E_n(0)$ is related to $E_{n-1}(0)$ by

$$E_n(0) = \min\{C + E_{n-1}(0),\ p_\Theta L + (1 - p_\Theta)E_{n-1}(0)\}, \tag{6.15}$$

$n = 2, 3, \ldots$. This expression can be obtained simply by substituting $E_n(0)$ for $E_2(0)$ and $E_{n-1}(0)$ for $E_1(0)$ in equation (6.14). Utilizing equations (6.13) and (6.15), the minimal total expected expense with climatological information can be determined for any desired number of occasions n.

For the situation in which a weather forecasting system is available to the decision maker, the stochastic dynamic programming recursion for $E_n(q), 0 < q \leq 1$, can be derived in an analogous fashion (Murphy et al., 1985). For completeness, this more complex expression is included in the Appendix (see equation 6.A1).

Figure 6.3 shows the structure of the optimal policy for three types of information: climatological (i.e., $q = 0$), imperfect forecasts (i.e., $0 < q < 1$), and perfect information (i.e., $q = 1$). When only climatological information is available to the decision maker, the structure of the optimal policy is of the following form:

(i) *do not protect* on the first $n - k_0$ occasions;
(ii) *protect* on the last k_0 occasions;

for some constant $0 \leq k_0 \leq n$ (see Figure 6.3a). Here k_0 is a function of the cost–loss ratio C/L and the climatological probability

(a) CLIMATOLOGICAL INFORMATION

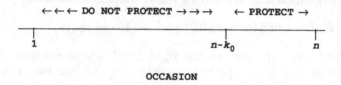

OCCASION

(b) PERFECT INFORMATION

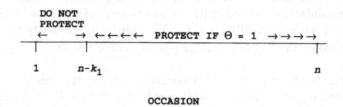

OCCASION

(c) IMPERFECT WEATHER FORECASTS

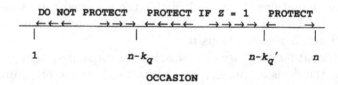

OCCASION

Figure 6.3. Structure of optimal policy for finite-horizon version of dynamic cost–loss ratio decision-making model for various forms of information: (a) climatological; (b) perfect; (c) imperfect weather forecasting system. (From Murphy et al., 1985)

of adverse weather p_Θ (see Krzysztofowicz and Long, 1990). This result is consistent with the phenomenon already observed for the two-occasion example, namely, that it is preferable to postpone protecting as long as possible, avoiding unnecessary protection in the event that the loss L is incurred.

Figure 6.3b shows the structure of the optimal policy given perfect information (i.e., $q = 1$):

(i) *do not protect* on the first $n - k_1$ occasions;

(ii) *protect* on the last k_1 occasions whenever $\Theta = 1$;

for some constant $k_1, 1 \leq k_1 \leq n$. Here k_1 again depends on the parameters C/L and p_Θ (see Krzysztofowicz and Long, 1990), and $k_1 \geq 1$ because it is always optimal to protect given $\Theta = 1$ in the static model (recall that $C/L < 1$). In this case of perfect information, the decision maker is assumed to be clairvoyant (but only concerning the next occasion, *not* the remaining periods), implying that protection needs be taken only on those occasions when it is known that adverse weather will occur. Hence, the structure of the optimal policy is of the same form as that for climatological information, except that protection can be initiated earlier (i.e., $k_0 \leq k_1$).

Figure 6.3c shows the structure of the optimal policy given the imperfect forecasting system (i.e., $0 < q < 1$):

(i) *do not protect* on the first $n - k_q$ occasions;

(ii) *protect* on occasions $n - k_q + 1$ through $n - k'_q$ whenever $Z = 1$;

(iii) *protect* on the last k'_q occasions;

for some constants k_q and $k'_q, 0 \leq k'_q \leq k_q \leq n$. The exact numerical values of k_q and k'_q depend on the parameters C/L and p_Θ, as well as on the forecast quality q (see Krzysztofowicz and Long, 1990). As in the case of climatological information, it is always optimal to protect on the last k'_q occasions. Moreover, as in the case of perfect information, it is optimal to protect given a forecast of adverse weather on the last k_q occasions. These constants are related by

$$k'_q \leq k_0 \leq k_q \leq k_1. \tag{6.16}$$

In particular, as the forecast quality q increases from zero to one, k_q increases from k_0 to k_1 and k'_q decreases from k_0 to zero. These results concerning the structure of the optimal policy follow directly from the corresponding stochastic dynamic programming recursion (see the Appendix and Krzysztofowicz and Long, 1990; Murphy et al., 1985).

Figure 6.4 shows some examples of the quality/value curves for the dynamic cost–loss ratio decision-making model when the cost $C = 0.3$, the loss $L = 1$, and the probability of adverse weather $p_\Theta = 0.2$, for several different lengths of finite horizon n, ranging from $n = 2$ to $n = 16$. Like the static (i.e., $n = 1$) model

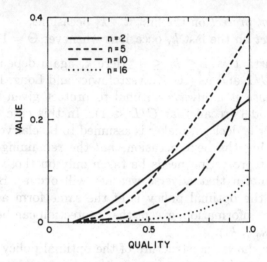

Figure 6.4. Quality/value curves for finite-horizon version of dynamic cost–loss ratio decision-making model, where $C = 0.3$, $L = 1$, and $p_\Theta = 0.2$, for $n = 2$, 5, 10, and 16. (From Murphy et al., 1985)

(see Figure 6.2), a quality threshold is still present below which the forecasting system has no economic value for the decision maker. Above this threshold, the economic value of the imperfect weather forecasting system increases in a piecewise linear fashion. The quality/value curve is convex, since the straight-line segments have increasing slope as quality q increases. The number of linear segments increases, resulting in curves that gradually become smoother in appearance, as the horizon n increases (Murphy et al., 1985).

3.3. Dynamic cost–loss ratio model (infinite horizon, discounted)

Although some decision-making applications, such as the fruit-frost problem already mentioned in Section 3.2, are inherently finite horizon, others involve a process that continues indefinitely into the future. For instance, the so-called fallowing/planting problem refers to the decision faced each year by a wheat farmer as to whether to plant a crop or to let the land lie fallow and accumulate soil moisture until planting time the following year (see Chapter 4 of this volume and Brown, Katz, and Murphy, 1986; Katz, Brown, and Murphy, 1987). It is reasonable to assume that such a decision maker wants to maximize the expected return totaled over a relatively large number of years (i.e., effectively, an *infinite* horizon).

Consequently, an infinite-horizon version of the dynamic cost–loss ratio decision-making model will now be treated. The nature of the dynamics and the form of imperfect weather forecasts remains the same as for the finite-horizon model previously considered (Section 3.2). However, when dealing with an economic optimization problem over a relatively long period of time, one additional complication does arise. Future expenses (or returns) need to be *discounted*, the rationale being that a dollar today is worth more than a dollar in the future because of uncertainties as to whether or not future expenses actually will be incurred and because of opportunities for investment and consumption. Specifically, an expense E incurred on the next occasion has a "present value" of only αE, where α, $0 < \alpha < 1$, is the *discount factor*. In relative terms, this "future value" E is diminished (or discounted) at the rate

$$r = \frac{E - \alpha E}{\alpha E} = \frac{1 - \alpha}{\alpha}, \qquad (6.17)$$

called the *discount rate* (for a more detailed explanation, see Alchian and Allen, 1972). In summary, a version of the dynamic cost–loss ratio model will be considered in which the decision maker minimizes expected expense, totaled and discounted (with discount factor α) over an infinite horizon.

In some respects, the infinite-horizon model is actually conceptually simpler to deal with than the finite-horizon model. Because an infinitely long future always remains, the decision-making process can be viewed, at least in a probabilistic sense, as starting over and over again. Therefore, it is straightforward to derive a stochastic dynamic programming recursion for $E(q)$, now representing the minimal expected expense, totaled and discounted over an infinite horizon. One might question why an infinite horizon is realistic when the loss L will surely be incurred within some finite number of occasions (unless protection is always taken). But it must be recognized that a positive probability always remains of not incurring the loss L over the first n occasions, no matter how large n is. Further, the infinite-horizon problem can often serve as a convenient approximation to the corresponding finite-horizon problem.

First, the situation is treated in which only climatological information is available to the decision maker. In this case, $E(0)$

satisfies the recursion

$$E(0) = \min\{C + \alpha E(0),\ p_\Theta L + (1 - p_\Theta)\alpha E(0)\}. \qquad (6.18)$$

The first (second) term within the brackets represents the total, discounted expected expense when protection is (is not) taken on the initial occasion. This recursion resembles the one for the finite-horizon situation depicted in equation (6.15), except that now the length of the horizon is always infinite (as opposed to finite length n or $n-1$) and the discount factor α has been introduced. Turning to the case in which the imperfect weather forecasting system is available to the decision maker, a recursion analogous to equation (6.18) can be derived for $E(q)$, $0 < q < 1$ (Katz and Murphy, 1990). This more complex expression is included in the Appendix (see equation 6.A2).

Using the recursion (6.18), Katz and Murphy (1990) showed that the optimal policy with climatological information alone is of the form:

(i) *protect if $p_\Theta > p_\Theta^*$;*
(ii) *do not protect if $p_\Theta < p_\Theta^*$.*

Here the threshold p_Θ^* for the climatological probability of adverse weather is given by

$$p_\Theta^* = \frac{C/L}{1 - [(C/L)/r]}. \qquad (6.19)$$

For this threshold to be well defined, expenses must be discounted at a fast enough rate (i.e., $\alpha < 1 - C/L$). Otherwise, protection is never optimal with only climatological information. The structure of the optimal policy resembles that for the static cost–loss ratio model (Section 3.1), with the discount rate r also entering into the threshold, depicted in equation (6.19). Similar, but more complex results can be obtained for the structure of the optimal policy when the imperfect forecasting system is available to the decision maker (Katz and Murphy, 1990).

Like the static and finite-horizon versions of the cost–loss ratio model, a threshold in quality exists below which the economic value of the forecasts is zero. This threshold arises because the optimal policy with the imperfect weather forecasting system does not deviate from that for climatological information alone, unless the quality q is sufficiently greater than zero. For example, suppose that $p_\Theta < p_\Theta^*$ (i.e., the optimal policy is not to protect for

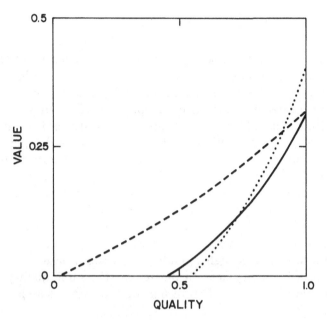

Figure 6.5. Quality/value curves for infinite-horizon version of dynamic cost–loss ratio decision-making model, where $\alpha = 0.9$: $C = 0.05$, $L = 1$, and $p_\Theta = 0.05$ (dashed line); $C = 0.2$, $L = 1$, and $p_\Theta = 0.2$ (solid line); and $C = 0.05$, $L = 1$, and $p_\Theta = 0.2$ (dotted line). (From Katz and Murphy, 1990)

climatological information). Then the quality threshold can be expressed as

$$q^* = (C/L) \left[\frac{1 - \alpha(1 - p_\Theta)}{1 - \alpha} \right] - p_\Theta \qquad (6.20)$$

(Katz and Murphy, 1990). This threshold is identical to the one for the static situation depicted in equation (6.9), except for the term involving the discount factor α. An analogous expression holds for the alternative case of $p_\Theta > p_\Theta^*$ (Katz and Murphy, 1990).

Above the quality threshold, it can be shown (Katz and Murphy, 1990) that the economic value of the imperfect weather forecasting system, $V(q)$, must increase at an increasing rate as the quality q increases (i.e., a convex curve). Figure 6.5 shows three examples of these quality/value curves. Their shape resembles the curves for the finite-horizon situation (Figure 6.4), but they are smooth rather than piecewise linear.

4. Extensions

4.1. Autocorrelated weather variables

All the results presented so far concerning the structure of the optimal policy and the shape of the quality/value curve have been
predicated upon the simplifying assumption that the states of the
weather variable are temporally independent. Of course, it is well
known that, depending on the time scale, most weather variables
may actually possess a substantial degree of temporal dependence,
usually exhibiting a tendency for a given weather state to persist
(i.e., positive autocorrelation). If the decision-making situation
being modeled were static (e.g., the static cost–loss ratio model
of Section 3.1), then the issue of autocorrelation could be treated
in a relatively straightforward manner. Knowing that a weather
variable is autocorrelated could be viewed simply as a special case
of having forecasts of higher quality than those based on the assumption of independence. In particular, the same quality/value
curve would apply (e.g., Figure 6.2 for the static cost–loss ratio
model).

With dynamic decision-making situations, however, the presence
of autocorrelation makes the problem inherently more complex.
For this reason, only a few attempts have been made to allow for
temporal dependence when assessing the economic value of imperfect weather forecasts (Epstein and Murphy, 1988; Katz, 1993;
Wilks, 1991). Nevertheless, given the recent interest in the economic value of forecasts of ENSO (El Niño–Southern Oscillation),
a phenomenon with substantial dependence on monthly, seasonal,
and even annual time scales, this issue is of practical importance
(Adams et al., 1995; Kite-Powell and Solow, 1994).

Katz (1993) has attempted to develop a more general analytical
framework in which forecasts of an autocorrelated weather variable could be treated naturally. This framework is briefly outlined.
Exactly the same prototype form of weather information introduced in Section 2.2 is retained. But now the sequence of weather
states (i.e., adverse or no adverse weather) $\{\Theta_t : t = 1, 2, \ldots\}$
is assumed to constitute a first-order Markov chain (e.g., Gabriel
and Neumann, 1962). Besides the probability of adverse weather
p_Θ, one new parameter needs to be incorporated into the model,
namely, the first-order autocorrelation or *persistence parameter*
$d = \mathrm{Corr}(\Theta_t, \Theta_{t+1})$. To allow for the persistence typical of weather

variables, we take $0 < d < 1$. The previous sections of this chapter have tacitly assumed that $d = 0$ (i.e., an independent climate). The other extreme of $d = 1$ represents a perfectly persistent climate.

A major complication for any dynamic decision-making problem arises because the assumed Markovian property implies that all future weather states are correlated with the present one; that is,

$$\mathrm{Corr}(\Theta_t, \Theta_{t+k}) = d^k, \ k = 1, 2, \ldots . \tag{6.21}$$

In effect, the decision maker has available on a given occasion a sequence of climatological "forecasts" for the indefinite future, whose skill cascades down to zero. In selecting the action for the present occasion, the decision maker should not ignore this information about the weather on future occasions. As an extreme example, if an orchardist knew for sure (or with a high probability) that the weather tomorrow night would destroy the fruit crop, then there would be no point in taking protective action tonight, no matter what the forecast for the present occasion.

The infinite-horizon, discounted version of the dynamic cost–loss ratio decision-making model, just treated in Section 3.3 for $d = 0$, is again considered, but now with an autocorrelated weather variable (i.e., $d > 0$). First, the situation in which only climatological information is available to the decision maker is discussed. A stochastic dynamic programming recursion that is a straightforward generalization of equation (6.18) can be derived (Katz, 1993). For a given strategy, the minimal expected expense, totaled and discounted over an infinite horizon, naturally depends on the persistence parameter d. Consequently, the optimal policy for climatological information with d sufficiently greater than zero might well differ from that for $d = 0$. A decision maker who naively believed that the states of the weather variable were temporally independent would in some circumstances actually be following a "suboptimal policy," incurring unnecessary additional expense. Katz (1993) described specific instances in which this phenomenon actually occurs.

With an autocorrelated weather variable, care must be taken in making any assumptions about the stochastic properties of the sequence of imperfect weather forecasts (introduced in Section 2.2). Recall that these forecasts are for the next occasion only, not for the remaining occasions. In this context, Z_t denotes a forecast of Θ_t available at time $t-1$. It is assumed that this sequence of forecast states $\{Z_t : t = 1, 2, \ldots\}$ has the same probabilistic structure as the weather variable, that is, a first-order Markov chain with the identical persistence parameter d. Because this requirement must hold in the two limiting situations of $d = 0$ and $d = 1$, it is a reasonable condition to impose in general.

It is necessary to make additional assumptions about the structure of the bivariate process of weather and forecast states $\{(\Theta_t, Z_t): t = 1, 2, \ldots\}$. Katz (1992) identified the exact conditions. The basic idea is that the present forecast (i.e., for the next occasion) should subsume any predictive information contained in past weather and forecast states, and that the next forecast (i.e., for the occasion after next) need only be based on the present weather state. In other words, the forecast variable can be thought of as "leading" the weather variable by one occasion.

As in the case of an independent weather variable, the quality of the imperfect weather forecasting system is specified through the conditional probability of adverse weather given a forecast of adverse weather

$$p_1 = \Pr\{\Theta_t = 1 | Z_t = 1\}, \text{ for } p_\Theta + d(1 - p_\Theta) \le p_1 \le 1. \quad (6.22)$$

Except that the time index t has been made explicit, equation (6.22) is identical to the original definition of p_1 in equation (6.2). The lower bound on p_1 is no longer p_Θ; instead, it reflects the temporal dependence of the weather variable, being $\Pr\{\Theta_{t+1} = 1 | \Theta_t = 1\}$. The measure of forecast quality q is defined in the same manner as previously in equation (6.6), with p_1 now specified by equation (6.22). This quality measure still represents the correlation between the forecast and weather variables [i.e., $q = \text{Corr}(Z_t, \Theta_t)$, $d \le q \le 1$]. The lower bound on q is d, rather than zero, because of the predictive capability of the Markov chain model itself.

Despite the fact that the weather forecasting system provides only one-occasion-ahead forecasts, all future weather states are correlated with the present forecast. In particular, it can be shown that the cross correlation function between the forecast and weather variables is of the form

$$\text{Corr}(Z_t, \Theta_{t+k-1}) = qd^{k-1}, \quad k = 2, 3, \ldots. \quad (6.23)$$

Here k denotes the lead time because Z_t is a forecast of Θ_t available at time $t - 1$. In other words, the combination of one-occasion-ahead forecasting quality and temporal dependence induces "forecasts" two occasions or more ahead, whose quality exceeds that based on the autocorrelation of the weather variable alone [i.e., $qd^{k-1} \geq d^k = \text{Corr}(\Theta_{t-1}, \Theta_{t+k-1})$, since $q \geq d$].

Because the requirement of overall calibration in equation (6.4) has been retained, the quality measure q still satisfies the sufficiency relation (see Section 2.2). Therefore, economic value $V(q)$ must be a nondecreasing function of q, $d \leq q \leq 1$. As already observed for the situation of climatological information alone, the presence of autocorrelation in a dynamic decision-making problem affects the expressions for the minimal expected expenses, totaled and discounted over an infinite horizon. Consequently, the economic value of a forecasting system with a fixed level of quality q might well differ depending on the degree of persistence d. Katz (1992) relied again on the approach of stochastic dynamic programming to determine these expenses.

Some numerical examples for the infinite-horizon, discounted version of the dynamic cost–loss ratio decision-making model are provided to illustrate how the shape and magnitude of the quality/value curve change as a function of the autocorrelation d. Figure 6.6 (top) shows an example in which the economic value of the imperfect weather forecasting system is relatively sensitive to the degree of autocorrelation of the weather variable. The quality/value curve remains convex for $d > 0$. Holding quality q constant, the economic value is substantially reduced as d increases, with the reduction being roughly linear in d. Moreover, the curve rises at a less rapid rate the greater the degree of persistence.

Figure 6.6 (bottom) relates to an example in which the economic value of the imperfect weather forecasting system is less sensitive to the degree of autocorrelation of the weather variable. Again, the curve is convex for $d > 0$. But the curve does not change much unless the degree of persistence is relatively high. Holding quality q constant, the reduction in economic value is a highly nonlinear function of d. On the other hand, the curve rises at nearly the same rate, no matter what the degree of persistence d.

It would be natural to make a further extension to the situation in which the weather forecasting system produces forecasts not only for the next occasion, but also for subsequent occasions. The quality of such forecasts would be assumed to decay with lead time, because of limits to predictability (see Chapter 1 of this volume). Since the situation just treated already involves, in effect, "forecasts" for subsequent occasions, this extension would be straightforward.

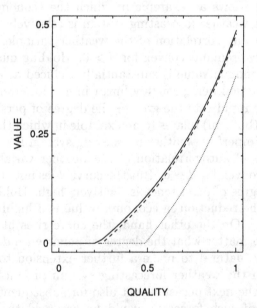

Figure 6.6. Quality/value curves for infinite-horizon version of dynamic cost–loss ratio decision-making model with an autocorrelated weather variable (solid line represents persistence parameter $d = 0$, dashed line represents $d = 0.25$, and dotted line represents $d = 0.5$): (top) $\alpha = 0.9$, $C = 0.15$, $L = 1$, and $p_\Theta = 0.2$; (bottom) $\alpha = 0.98$, $C = 0.01$, $L = 1$, and $p_\Theta = 0.025$. (From Katz, 1992)

4.2. Other extensions

There are several other respects in which one could attempt to generalize upon the results presented in this chapter. For instance, Katz (1987) examined the implications for the shape of the quality/value curve of relaxing the requirement of overall reliability of the imperfect weather forecasting system in equation (6.4). The constraint (6.5) requires that the conditional probability of adverse weather p_0 move from the climatological probability of adverse weather p_Θ toward zero at the same rate at which the other conditional probability p_1 moves from p_Θ toward one (see Figure 6.1d). Instead, these two parameters, p_0 and p_1, are now free to move toward their respective limits at rates that are independent of one another. The sufficiency condition (6.3) is satisfied, and economic value must be a nondecreasing function of p_1 and a nonincreasing function of p_0. We do not treat the most general situation in which p_0 might increase as p_1 decreases or vice versa, because the sufficiency condition would not be satisfied.

Because the original definition of the measure of forecast quality, equation (6.6), is predicated upon the overall reliability condition (6.4) being in force, first this measure needs to be extended to the more general situation now being considered. The quality of the forecasting system now depends on both p_0 and p_1, and it can be completely characterized only by a two-dimensional measure. Nevertheless, one natural way to generalize the forecast quality measure is to retain its property of being the correlation between the forecast and weather variables [i.e., $q = \mathrm{Corr}(Z, \Theta)$]. In general, this correlation coefficient is given by

$$q = \left[\frac{(p_\Theta - p_0)(p_1 - p_\Theta)}{p_\Theta(1 - p_\Theta)} \right]^{1/2}. \tag{6.24}$$

The generalized measure of quality q depends explicitly on both p_0 and p_1, reflecting the distance of these two parameters from p_Θ. It is easy to show that the original quality measure (6.6) is obtained in the special case in which the constraint (6.5) is substituted into equation (6.24). Of course, the generalized measure still has the inherent limitation of being one-dimensional.

Katz (1987) studied the shape of the quality/value curve for the static cost–loss ratio decision-making model (Section 3.1) under these weaker conditions being imposed on the weather forecasting system. If no assumptions are made about the relative rates at which p_0 and p_1 move toward their respective limits, then the shape of the quality/value curve cannot be characterized in a simple manner. All that is guaranteed is that economic value remains a nondecreasing function of q, because the sufficiency condition (6.3) is still satisfied. In particular, its shape is no longer necessarily convex, but may be concave or even locally convex for certain ranges of forecast quality and locally concave for other ranges of quality. Of course, it could be argued that the constraint (6.5) represents a plausible way for improvements in a weather forecasting system to be realized in practice.

As part of their study of the finite-horizon version of the dynamic cost–loss ratio decision-making model (Section 3.2), Krzysztofowicz and Long (1990) also dealt with a more complex form of weather forecast information than the prototype form treated in this chapter (Section 2.2). In addition to the case of a two-state forecast variable, the situation is analyzed in which probability forecasts [i.e., a continuous variable on the interval $(0,1)$] are available to the decision maker. By making use of a particular form of parametric model (i.e., the beta distribution), analytical solutions for the structure of the optimal policy can be derived for this more realistic form of forecast information.

The structure of the cost–loss ratio decision-making model (Section 3.1) also can be generalized in several respects. For instance, it would be natural to allow for more than two actions and more than two states of weather (Murphy, 1985). Nevertheless, it is difficult to obtain tractable, analytical results concerning the structure of the optimal policy and the shape of the quality/value curve in such situations. Perhaps, a more conceptually appealing approach would be to generalize the model directly to the situation in which the weather variable is continuous. In particular, a starting point could be the case in which the joint distribution of the forecast and weather variables is bivariate normal. Such an assumption is reasonable for temperature and was employed in the fruit-frost problem (Katz et al., 1982). One could allow for only two possible actions, or also permit the action to be a continuous variable (e.g., Gandin, Murphy, and Zhukovsky, 1992). A treatment of a continuous weather variable that allows for autocorrelation could follow Krzysztofowicz (1985).

5. Implications

The relationship between the scientific quality and economic value of imperfect weather forecasts has been examined for various forms of prototype decision-making models. Although all these models are simpler than most real-world decision-making problems, they do retain some essential features of such situations, including the fact that many of these situations are dynamic in nature. A virtually ubiquitous result is the convex shape of the quality/value curve. Economic value is zero for forecasting systems whose quality falls below a threshold. Above this threshold, economic value rises at an increasing rate as forecast quality increases toward that of perfect information. A quality threshold also arises in the fallowing/planting problem (Brown et al., 1986). However, neither this problem nor the fruit-frost problem (Katz et al., 1982) necessarily possess a quality/value curve whose shape is convex.

These results based on prototype decision-making models have some important implications for research both on weather forecasting and on the economic value of forecasts. In particular, the existence of a quality threshold may explain why current long-range (i.e., monthly or seasonal) weather forecasts, which are necessarily of relatively low quality, are apparently ignored by many decision makers (Changnon, Changnon, and Changnon, 1995; Easterling, 1986; also see Chapter 5 of this volume). Moreover, the convexity of the quality/value curve is somewhat discouraging with respect to the potential benefits of realistic improvements in the quality of weather forecasts, at least for those cases in which forecast quality is now relatively far from that of perfect information. However, it must be kept in mind that the rate at which economic value actually increases will depend on the rate at which quality improves, and quality might well be a concave function of monetary investment in meteorological research.

The prototype forms of decision-making models that have been treated were motivated in part by case studies of real-world decision-making situations, such as the fruit-frost and fallowing/planting problems. A vital need exists for more such case studies of the economic value of imperfect weather and climate forecasts in real-world applications (as observed in Chapter 4 of this volume). Studies of the analytical properties of prototype decision-making models like those presented in this chapter play an important complementary role. Specifically, they help to distinguish those features of case studies that are truly novel from others that ought to have been anticipated from theoretical value-of-information studies based on decision making under uncertainty.

Appendix

Stochastic dynamic programming

Finite-horizon dynamic cost–loss ratio model. The stochastic dynamic programming recursion for the minimal total expected expense $E_n(q)$ for the imperfect weather forecasting system is

$$E_n(q) = p_\Theta \min\{C + E_{n-1}(q), \; p_1 L + (1 - p_1) E_{n-1}(q)\}$$
$$+ \; (1 - p_\Theta) \min\{C + E_{n-1}(q), \; p_0 L + (1 - p_0) E_{n-1}(q)\}, \tag{6.A1}$$

$n = 1, 2, \ldots$; $0 \le q \le 1$, with the convention that $E_0(q) = 0$ (Murphy et al., 1985). The conditional probabilities, p_0 and p_1, that appear in equation (6.A1) can be expressed as functions of the forecast quality q through equations (6.5) and (6.6). The first term in curly brackets on the right-hand side of equation (6.A1) represents the minimal total expected expense over the n occasions when adverse weather is forecast (i.e., $Z = 1$) on the first of the n occasions, whereas the second term in curly brackets represents the corresponding expense for the case of a forecast of no adverse weather (i.e., $Z = 0$) on the first occasion.

Infinite-horizon, discounted dynamic cost–loss ratio model. The stochastic dynamic programming recursion for the total, discounted expected expense $E(q)$ for the imperfect weather forecasting system is

$$E(q) = p_\Theta \min\{C + \alpha E(q), \; p_1 L + (1 - p_1) \alpha E(q)\}$$
$$+ \; (1 - p_\Theta) \min\{C + \alpha E(q), \; p_0 L + (1 - p_0) \alpha E(q)\}, \tag{6.A2}$$

$0 \le q \le 1$ (Katz and Murphy, 1990). Again, p_0 and p_1 are related to q through equations (6.5) and (6.6). The first term in curly brackets on the right-hand side of equation (6.A2) represents the total discounted expected expense given a forecast of adverse weather (i.e., $Z = 1$) on the initial occasion, whereas the second term in curly brackets represents the corresponding expense for the case of a forecast of no adverse weather (i.e., $Z = 0$) on the initial occasion.

Acknowledgments

We thank Roman Krzysztofowicz and Daniel Wilks for comments. This chapter summarizes research that was supported in part by the National Science Foundation under grants ATM-8714108 and SES-9106440.

References

Adams, R.M., Bryant, K.S., McCarl, B.A., Legler, D.M., O'Brien, J., Solow, A. & Weiher, R. (1995). Value of improved long-range weather information. *Contemporary Economic Policy*, **XIII**, 10–19.

Alchian, A.A. & Allen, W.R. (1972). *University Economics: Elements of Inquiry* (third edition). Belmont, CA: Wadsworth.

Baquet, A.E., Halter, A.N. & Conklin, F.S. (1976). The value of frost forecasting: a Bayesian appraisal. *American Journal of Agricultural Economics*, **58**, 511–520.

Blackwell, D. (1953). Equivalent comparisons of experiments. *Annals of Mathematical Statistics*, **24**, 265–272.

Brown, B.G., Katz, R.W. & Murphy, A.H. (1986). On the economic value of seasonal-precipitation forecasts: the fallowing/planting problem. *Bulletin of the American Meteorological Society*, **67**, 833–841.

Changnon, S.A., Changnon, J.M. & Changnon, D. (1995). Uses and applications of climate forecasts for power utilities. *Bulletin of the American Meteorological Society*, **76**, 711–720.

DeGroot, M.H. (1970). *Optimal Statistical Decisions*. New York: McGraw-Hill.

Easterling, W.E. (1986). Subscribers to the NOAA *Monthly and Seasonal Weather Outlook*. *Bulletin of the American Meteorological Society*, **67**, 402–410.

Ehrendorfer, M. & Murphy, A.H. (1988). Comparative evaluation of weather forecasting systems: sufficiency, quality, and accuracy. *Monthly Weather Review*, **116**, 1757–1770.

Ehrendorfer, M. & Murphy, A.H. (1992a). Evaluation of prototypical climate forecasts: the sufficiency relation. *Journal of Climate*, **5**, 876–887.

Ehrendorfer, M. & Murphy, A.H. (1992b). On the relationship between the quality and value of weather and climate forecasting systems. *Időjárás*, **96**, 187–206.

Epstein, E.S. & Murphy, A.H. (1988). Use and value of multiple-period forecasts in a dynamic model of the cost–loss ratio situation. *Monthly Weather Review*, **116**, 746–761.

Gabriel, K.R. & Neumann, J. (1962). A Markov chain model for daily rainfall occurrence at Tel Aviv. *Quarterly Journal of the Royal Meteorological Society*, **88**, 90–95.

Gandin, L.S., Murphy, A.H. & Zhukovsky, E.E. (1992). Economically optimal decisions and the value of meteorological information. *Preprints, Fifth International Meeting on Statistical Climatology*, J64–J71. Toronto: Atmospheric Environment Service.

Hilton, R.W. (1981). The determinants of information value: synthesizing some general results. *Management Science*, **27**, 57–64.

Katz, R.W. (1987). On the convexity of quality/value relations for imperfect information about weather or climate. *Preprints, Tenth Conference on Probability and Statistics in Atmospheric Sciences*, 91–94. Boston: American Meteorological Society.

Katz, R.W. (1992). Quality/value relationships for forecasts of an autocorrelated climate variable. *Preprints, Fifth International Meeting on Statistical Climatology*, J91–J95. Toronto: Atmospheric Environment Service.

Katz, R.W. (1993). Dynamic cost–loss ratio decision-making model with an autocorrelated climate variable. *Journal of Climate*, **5**, 151–160.

Katz, R.W., Brown, B.G. & Murphy, A.H. (1987). Decision-analytic assessment of the economic value of weather forecasts: the fallowing/planting problem. *Journal of Forecasting*, **6**, 77–89.

Katz, R.W. & Murphy, A.H. (1987). Quality/value relationship for imperfect information in the umbrella problem. *The American Statistician*, **41**, 187–189.

Katz, R.W. & Murphy, A.H. (1990). Quality/value relationships for imperfect weather forecasts in a prototype multistage decision-making model. *Journal of Forecasting*, **9**, 75–86.

Katz, R.W., Murphy, A.H. & Winkler, R.L. (1982). Assessing the value of frost forecasts to orchardists: a dynamic decision-making approach. *Journal of Applied Meteorology*, **21**, 518–531.

Kite-Powell, H.L. & Solow, A.R. (1994). A Bayesian approach to estimating benefits of improved forecasts. *Meteorological Applications*, **1**, 351–354.

Krzysztofowicz, R. (1985). Bayesian models of forecasted time series. *Water Resources Bulletin*, **21**, 805–814.

Krzysztofowicz, R. & Long, D. (1990). To protect or not to protect: Bayes decisions with forecasts. *European Journal of Operational Research*, **44**, 319–330.

Krzysztofowicz, R. & Long, D. (1991). Forecast sufficiency characteristic: construction and application. *International Journal of Forecasting*, **7**, 39–45.

Murphy, A.H. (1977). The value of climatological, categorical and probabilistic forecasts in the cost–loss ratio situation. *Monthly Weather Review*, **105**, 803–816.

Murphy, A.H. (1985). Decision making and the value of forecasts in a generalized model of the cost–loss ratio situation. *Monthly Weather Review*, **113**, 362–369.

Murphy, A.H. & Ehrendorfer, M. (1987). On the relationship between the accuracy and value of forecasts in the cost–loss ratio situation. *Weather and Forecasting*, **2**, 243–251.

Murphy, A.H., Katz, R.W., Winkler, R.L. & Hsu, W.-R. (1985). Repetitive decision making and the value of forecasts in the cost–loss ratio situation: a dynamic model. *Monthly Weather Review*, **113**, 801–813.

Ross, S.M. (1983). *Introduction to Stochastic Dynamic Programming*. New York: Academic Press.

Thompson, J.C. (1952). On the operational deficiencies in categorical weather forecasts. *Bulletin of the American Meteorological Society*, **33**, 223–226.

White, D.J. (1966). Forecasts and decisionmaking. *Journal of Mathematical Analysis and Applications,* **14**, 163–173.

White, D.J. (1978). *Finite Dynamic Programming: An Approach to Finite Markov Decision Processes.* Chichester, UK: Wiley.

Wilks, D.S. (1991). Representing serial correlation of meteorological events and forecasts in dynamic decision-analytic models. *Monthly Weather Review,* **119**, 1640–1662.

Winkler, R.L. & Murphy, A.H. (1985). Decision analysis. In *Probability, Statistics, and Decision Making in the Atmospheric Sciences,* ed. A.H. Murphy & R.W. Katz, 493–524. Boulder, CO: Westview Press.

Index

Printed in the United States
By Bookmasters